Modern Irrigation Technologies for Smallholders in Developing Countries

GEZ CORNISH

Practical ACTION PUBLISHING

Practical Action Publishing Ltd
25 Albert Street, Rugby, CV21 2SD, Warwickshire, UK
www.practicalactionpublishing.org

© HR Wallingford Group Ltd, 1998

First published 1998\Digitised 2008

ISBN 13 Paperback: 978 1 85339 457 7

ISBN Library Ebook: 9781780444178
Book DOI: http://dx.doi.org/10.3362/9781780444178

Since 1974, Practical Action Publishing has published and disseminated books
and information in support of international development work throughout
the world. Practical Action Publishing is a trading name of Practical Action
Publishing Ltd (Company Reg. No. 1159018), the wholly owned publishing
company of Practical Action. Practical Action Publishing trades only in support
of its parent charity objectives and any profits are covenanted back to Practical
Action (Charity Reg. No. 247257, Group VAT Registration No. 880 9924 76).

Funding for the preparation of this work was provided by the Department for
International Development (DFID)

Summary

The objective of this report, commissioned by the Engineering Division of DFID, is to identify pre-conditions relating to water availability, institutional support and economic opportunity that must be satisfied before smallholders in developing countries can adopt modern irrigation methods. The report also reviews the range of irrigation hardware that is available and indicates the types of equipment that are more likely to meet the requirements of the smallholder sector.

A broad definition of the term "smallholder" is adopted in the report. The term describes farmers practising a mix of commercial and subsistence production where the family provides the majority of labour and the farm provides the principal source of income. It also includes small commercial enterprises growing high value crops such as cut flowers and produce for export. A smallholder will normally derive his/her livelihood from an irrigated holding of less than 2 to 5 ha - holdings are often less than 0.2 ha. Larger enterprises often have greater access to assistance in design, operation and marketing and the findings of this report may be less relevant to these farm types. The report addresses both individual farmers and smallholder schemes where many farmers share some part of the water supply infrastructure.

The report is divided into the following chapters.

Chapter 1: Sets out the potential role of modern irrigation methods against the background of increasing water scarcity and continuing food shortages, particularly in sub-Saharan Africa. The potential shortcomings of introducing technologies developed in other environments, as a means of improving agricultural productivity and the livelihood of smallholder farmers, are set out and the need to draw lessons from past successes and failures is underscored.

Chapter 2: Describes and classifies the range of modern irrigation technologies and considers the characteristics of those technologies making them more or less suited for use by smallholders.

Chapter 3: Defines the technological characteristics required of equipment that will be used by smallholders.

Chapter 4: Reviews the experience of smallholders with modern irrigation technologies in a range of economic and agro-ecological conditions. It aims to summarise the conditions faced by smallholders that determine their willingness to adopt and maintain modern irrigation technologies. Information is presented from eleven countries as diverse as Israel, India, Zimbabwe and Guatemala. These are countries where information on the use of modern irrigation methods by smallholders is reported in the literature or where the author has had direct experience. The diversity of conditions seen amongst these countries assists in identifying what are the essential and common elements where modern technologies have been adopted.

Israel and Cyprus in particular stand out as high income and upper-middle income economies in a report focused on developing countries. These countries have been included because much of their irrigated agriculture takes place on a small scale with farmers' landholdings generally less than 2 ha. Farmers in these two countries share few of the characteristics of smallholders in poorer developing countries. However, the experience of these countries in their adoption of modern methods and the role of major public sector investment in water conveyance and distribution to supply individual smallholders merits their inclusion.

Chapter 5: Briefly examines the potential for modern irrigation technologies on smallholder schemes in Africa.

Chapter 6: Summarises the findings of the study drawing conclusions regarding both the types of equipment that are likely to be appropriate and the wider economic, social and policy issues that must be in place before smallholders are likely to exploit the potential benefits of modern technologies.

Chapter 7: Outlines the issues that must be addressed to promote the use of modern irrigation methods in developing countries.

The report contains an extensive bibliography on the theme of modern irrigation methods and their adoption by smallholder farmers. Details of the country experience summarised in Chapter 4 are contained in Appendix 2.

Contents

Contents continued

Figures

Appendices

1 Introduction

General

The irrigated agriculture sector, which currently accounts for two-thirds of the world's water use, is increasingly required to produce more food from a limited land area using less water. Over the period 1979-1984, population growth outstripped food production in 24 African countries (World Bank, 1994). The Food and Agriculture Organisation (FAO, 1996) emphasised the importance of water for achieving food security. However, water resources are increasingly being exhausted, and competition for the available water between agriculture and the municipal and industrial sectors is increasing each year.

In 1990, the global irrigated area was estimated at 255 million hectares (Field 1990), of which some 200 million hectares were in low or middle income economies according to the classification of the World Bank (Figure 1). A high proportion of the land is farmed by smallholders. By far the greatest part, 140 million hectares, is in South and South East Asia. In this region, rice is the dominant irrigated crop, accounting for some 100 million hectares or 70% of area under irrigated cultivation. This area of land is clearly not suited for modern irrigation methods without major change in farmers' cropping preference. In Africa, some degree of water management is practised on some 14 million hectares, of which over 2 million are planted to rice (FAO, 1995 b). Thus, excluding areas growing rice, traditional surface methods are practised on a total area of over 50 million hectares in South and South East Asia and Africa. As water becomes increasingly scarce it will become necessary to convert at least some of this area to irrigation by modern systems. However, particular potential appears to lie in areas not yet developed for irrigation.

Appropriate modern irrigation methods, suited to the needs of the smallholder, potentially offer considerable scope for saving water, increasing production, and improving well-being in Africa. The continent includes 13 of the 18 nations of the world having less than 1,000 m^3/head/year of water, a situation of "absolute water scarcity". Regional food shortages are a constant threat and water shortage can only increase. Yet, over the African region as a whole, 35 million hectares of potentially irrigable land remain to be developed (FAO, 1995 b).

New approaches to irrigation development, which move away from the large-scale, top-down schemes of the 1970s with associated high costs and disappointing performance, are now sought. There is a focus in the world community on simple appropriate technologies supported by private sector investment, particularly for low-lift pumping, the exploitation of shallow aquifers and irrigation in peri-urban areas (FAO, 1995 a).

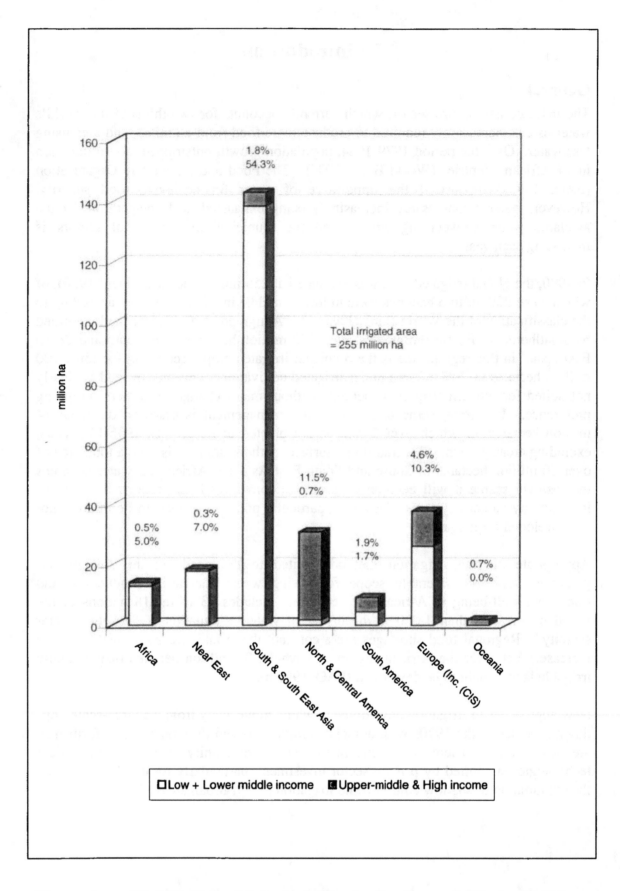

Figure 1   Distribution of Irrigated Area by Region and Economy (After Field 1990)

Modern Technologies

It is maintained by some within the irrigation community (Or, 1993), that modern irrigation technologies[1] are the key to increased food production and improvements in water use efficiency in the developing world. They advocate the wide-spread adoption of pressurised distribution networks and modern in-field irrigation systems as a response to regional shortages of food, land and water. The technical benefits of modern systems, by comparison with surface irrigation methods, are claimed to be:

- Improved conveyance and application efficiency, leading to a saving of water and a reduced risk of raised water tables
- Improved control over the timing and depth of irrigation, leading to possible improvements in yield and quality of output
- Reduced demand for labour
- Effective irrigation of coarse or shallow soils and sloping lands
- Better use of small discharges
- Reduction in the land taken up by the distribution system
- Better use of poor quality water, provided appropriate management practices are adopted
- Reduced risk to health by elimination of standing water.

Despite such apparent benefits, the use of modern methods is still largely confined to commercial, high-input agriculture, mostly in the developed world.

Development of sprinkler technology has been directed towards the needs and operating conditions typical of large-scale farming systems in North America and Western Europe. The technology is designed to lower the total costs of irrigation by reducing the requirements for labour and energy.

Micro-irrigation systems were developed in Israel, driven by the need to save water and a desire to expand agricultural production on to marginal desert soils using poor quality water. Equipment has become increasingly sophisticated over time. As with sprinkler technologies, the aim has been to improve application efficiency and uniformity, reduce labour inputs and lower the cost of installation on larger field areas.

Projects introducing modern irrigation technologies in the developing world have often failed. There is a clear danger of mis-matching irrigation hardware, developed for one set of physical and socio-economic conditions, with the circumstances in an entirely different environment. The resources available to operate and maintain equipment under high-input commercial agriculture bear no resemblance to those available to a smallholder. None the less, there is some common ground between large-scale commercial farmers and smallholders in the need to reduce labour costs and efforts, minimise energy costs and increase application efficiency and uniformity, but the relative value placed on these objectives will vary.

---

[1] For the purposes of this report, the term modern irrigation technology (or method) refers to any system involving pressurised distribution of water by pipeline at farm or field level.

In view of the renewed interest shown by national governments, international agencies and donors in exploiting the potential of modern irrigation methods on small-scale developments, it is important to analyse past successes and failures so as to guide the design of future projects. The present document draws from the experience of a number of countries. The circumstances in which modern technologies were introduced are identified, and the relative success or otherwise of the initiatives are summarised. Technologies now available vary widely. To help planners identify appropriate choices, a brief review of equipment is included, focused primarily on aspects which make the system more, or less, suited to smallholder farms in the developing world. Potential applications within the African region are reviewed in Chapter 5 and overall conclusions are presented in Chapters 6 and 7.

2 Modern Irrigation Technologies

The Development of Modern Irrigation Technologies

For the present purposes a modern irrigation technology is considered to be any irrigation system using piped distribution under pressurised or gravity head, at the farm or field level. Sophisticated control systems for surface irrigation are not considered.

The first sprinkler systems were developed in the USA to apply water in the field more efficiently than was possible with surface irrigation and to eliminate the need for labourers to tend and adjust flows in the field continually. Subsequent developments of overhead sprinkler technologies were driven primarily by the need to reduce labour requirements further. Large-scale machines were developed to reduce investment costs per hectare and meet the requirements of farmers in the USA and France, where irrigated agriculture is extensive and land holdings are large. More recently, as energy costs increased relative to other operating costs, the trend has been to reduce the operating pressure of sprinkler systems. Where this has led to higher field application efficiencies - as seen in the LEPA systems (see page 11) - this was normally a secondary objective of the developers.

Prior to 1960 most of Israel's horticultural and orchard crops were irrigated by solid-set or portable sprinkler systems. Simple drip systems were developed and evaluated to expand agricultural production on to marginal desert soils using poor quality water. They were shown to enable crop production on marginal soils, with poor quality water and with higher water use efficiencies than sprinkler irrigation.

Melamed (1989) states that the following factors led to the development of micro-irrigation and its subsequent widespread adoption by Israeli farmers:

- Water scarcity, leading to high costs for water
- Desert soils and poor water quality
- Requirement to avoid wetting of foliage with brackish water
- Skilled agricultural labour force
- Effective extension agency with a dedicated Irrigation and Soils Field Service
- Hardware manufacturers in close contact with farmers through the Kibbutzim.

Some of these circumstances may be replicated in developing countries but others, in particular the presence of a skilled labour force and a dedicated extension agency, are unlikely to be found.

Micro-irrigation hardware has become increasingly sophisticated over time in attempts to overcome operational problems, improve application uniformity, facilitate system design and reduce labour requirements to install and operate systems. Developments include:

- Improved media and mesh filters
- Chemigation systems
- Pressure-compensating emitters
- Self-flushing emitters

- Automation of system control
- Improved formulation of plastics to improve durability of components
- Light-weight, single season, drip lines

Many of the developments were intended to reduce labour inputs and permit lower cost installations on larger field areas, whilst maintaining potentially very high field application efficiencies.

Characteristics of Modern Irrigation Technologies

There is a wide variety of modern irrigation technologies ranging from large-scale irrigation machines such as centre pivots and linear move systems, to draghose sprinkler systems, drip and mini-sprinkler systems and simple piped distribution networks. Figure 2 shows a classification of modern irrigation technologies following the general classes of technology adopted by Rolland (1982), Keller and Bliesner (1990) and Hlavek (1995).

The primary division is between sprinkler, micro-irrigation and piped distribution systems for surface irrigation. The classification is based on the method by which water is applied and the fraction of the field area that is wetted. *Sprinkler technologies* spray water above the crop canopy, wetting the entire field area. *Micro-irrigation* applies water to only a fraction of the field surface, with water delivered through a network of pipes on to the soil directly or via small spray or bubbler (controlled orifice) outlets. Care is needed to avoid confusion between the terms 'modern irrigation technology' and 'micro-irrigation technology'; the latter being only one aspect of the former. *Piped distribution systems* are relatively simple in conception. Buried or surface pipe networks, which may be permanent or portable, replace open conveyance channels between the water source and field plots. Water is still applied to the plot by conventional means.

Centre pivot machines fitted with low pressure drop hoses to deliver water below the crop canopy (Low-Energy Precision-Application, LEPA) combine aspects of sprinkler and micro-irrigation technologies. Water is delivered at low pressure from small orifice nozzles, wetting only a small surface area. These systems are classified as a type of continuous move sprinkler system as they rely on centre pivot or linear move technology for their operation and maintenance.

Many of the technologies will not be suitable for the smallholder. The purpose of the present chapter is to define what is available on the market and to characterise the product so as to identify features both suitable and unsuitable for application on small farms in the developing world. More detailed information on the different technologies is given in Rolland (1982) and Keller and Bliesner (1990). A summary of the advantages, disadvantages, purchase costs and labour requirements of each system type is given in Table 1 at the end of this chapter.

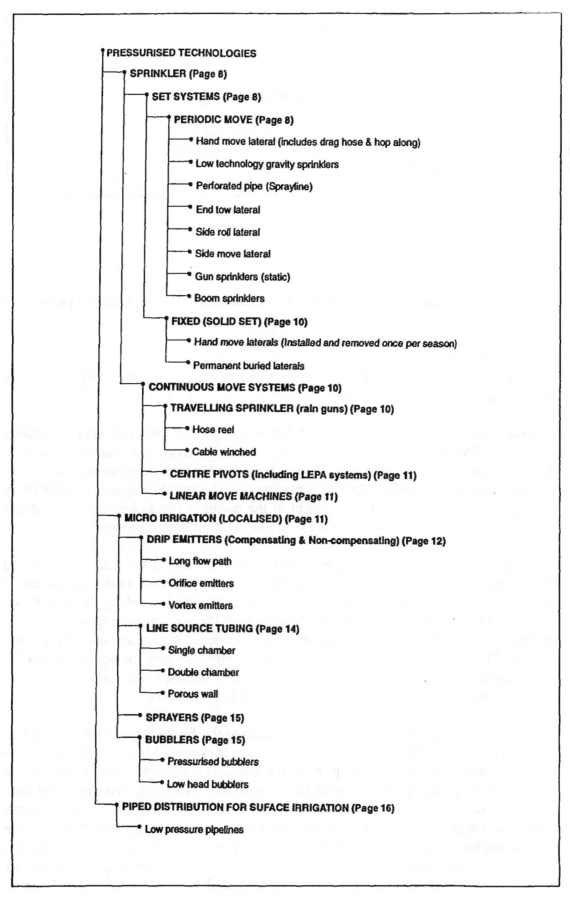

Figure 2   Classification of Modern Irrigation Technologies

Sprinkler Systems

Set Sprinkler Systems

Sprinkler technologies are divided into set systems and continuous move systems. In set systems the watering unit does not move whilst irrigating.

Set Systems - Periodic Move

Set systems that require the periodic movement of pipe work and/or sprinklers are classified as periodic move systems.

1. *Hand-move lateral* - Laterals are of lightweight aluminium with quick coupling connectors. The laterals are laid on the surface and fed from hydrants on a mainline, which may be buried or portable. Pipes are light but robust to withstand regular movement and re-assembly. Operating pressures at the sprinkler head generally lie in the range 200 - 400 kPa. Hand movement of the complete lateral between sets reduces capital outlay on equipment but results in very high labour requirements.

Hopalong systems use a bayonet-type valved coupler to connect sprinkler standpipes to laterals making it possible to move sprinklers between positions whilst the system is still operating. Sprinklers are placed in alternate riser positions along the lateral and then moved to the second position half-way through the set. By having two sprinkler 'set locations' for each position of the lateral, the diameter, weight and cost of pipes are reduced. The laterals do not need to be moved so often because the equipment can be left to operate through the night. A lateral is normally moved in the early morning and sprinklers are moved to their second set position in late afternoon.

Drag hoses offer an alternative method of reducing the frequency of lateral moves to cover a given area. Sprinklers are connected to the lateral via a hose equal in length to the normal lateral spacing. Each sprinkler thereby covers three positions from a single lateral location reducing the number of lateral moves by two-thirds. However, the time required to move laterals between positions is greater than in hand-moved systems owing to the need to disconnect and reconnect the hoses. Zadrazil (1990) describes a design using longer hoses and semi-permanent laterals moved only at the beginning and end of the season. Irrigation of the entire area is achieved by moving each sprinkler between five set positions, using a hose length equivalent to twice the 'normal' lateral spacing.

2. *Gravity fed artisan sprinklers* - In a number of developing regions of the world, sprinklers are produced locally for use mainly on systems developing gravity head. Manufacturing standards are lower than for internationally marketed equipment. Application efficiency and uniformity are also likely to be lower but the products meet the needs of local farmers. Bedini (1995) provides a comprehensive description of such sprinklers used in Kenya.

3. *Perforated pipe or sprayline* -Traditionally used for horticultural crops and in plant nurseries. Pipe diameters lie in the range 50 - 100 mm. The pipe may be stationary or driven by an oscillating mechanism. Small diameter (1 - 2 mm) holes are drilled directly into the lateral pipe, or screw-in jets may be used. Operating pressure varies between 40 and 200 kPa. Small droplets cover a wetted strip of between 7 to 15 m, depending on pressure. Application rates are relatively high, varying from 10 - 30 mm/hr. High labour input is needed to move sprayline pipes between settings.

4. *End tow lateral* - By using rigid pipe couplings and mounting the lateral on skid plates or small wheels it is possible to tow a lateral between consecutive positions on alternate sides of a centrally positioned main line using a small tractor. Labour requirements are greatly reduced when compared with hand-move systems. The system is not suited for use in small or irregular fields or on steep or uneven terrain. Care must be taken when towing the pipe to avoid sharp turns. Relatively skilled labour is needed.

5. *Side roll lateral* - The lateral is supported above the crop on large diameter wheels, the lateral forming the axle between the wheels. Lateral diameter is normally 100 or 125 mm. A small engine, mounted at the centre point of the lateral, is used to move the lateral between set positions. The length of lateral is normally about 400 m. Sprinklers are mounted on to the lateral using swivel couplings allowing the lateral to rotate whilst the sprinkler remains upright. The system is suitable for low crops in large, unobstructed rectangular fields of uniform slope.

6. *Side move lateral* - The lateral is carried above the crop on wheeled A-frames, the height of the frame being dependent on the crop type. The wheels on the A-frames can be swivelled through 90 degrees allowing the assembly to move in the direction of the pipe axis. Sprinklers may be mounted directly on the lateral or on small diameter (32 - 38 mm) aluminium pipes, up to 100 m long, trailed from the lateral for low-growing crops. A small engine, mounted at the mid point of the lateral, drives the wheels through a drive shaft and variable diameter pulleys which maintain the straight alignment of the lateral. Like the side roll system, these machines are designed for use in large, rectangular, flat fields that are free of obstructions. Side move systems require good technical and workshop facilities owing to their complexity.

7. *Static gun sprinklers* - Gun sprinklers operate at pressures between 400 to 700 kPa with a wetted diameter up to 100 m. Flow rate is between 8 and 30 l/s. Application rates tend to be high, and large droplets can damage soil structure. Static guns may be supplied by flexible hose or aluminium pipe. The gun may be moved manually or may be tractor towed between positions. Gun sprinklers are well suited to supplementary irrigation and can be used on small, irregular shaped fields. Static guns have been widely superseded by continuous travelling rain guns that greatly reduce labour requirements.

8. *Boom sprinklers* - Like gun sprinklers, boom sprinklers distribute water over a large wetted area. Wetted diameter is typically in the range 110 - 170 m depending on boom length. Operating pressures are between 500 and 800 kPa and application rates may be as low as 5 mm/hr. Field application rates are lower and droplet size is smaller, so there is less damage to soil structure than with a gun. Booms may be self-propelled or tractor-hauled between positions. They have a number of disadvantages. Uniformity of water application is highly sensitive to pressure fluctuations and wind effects, which can halt the boom's rotation. Choice of boom length is determined by field size. Several different machines may be needed where field sizes differ. It may be difficult to move the machine, particularly when the soil is wet. Large boom sprinklers have not found wide acceptance amongst commercial farmers.

Set Systems - Fixed or Solid Set

In fixed or solid set systems the reduced cost of labour to move equipment must be set against higher capital expenditure on more equipment. Water is not normally applied simultaneously over the entire field area as large capacity pumps and mains would be required. Different parts of the field are irrigated in sequence using manual or automatic control valves. For annual crops, laterals, sprinklers and possibly sub-mains are laid out on the surface after planting and removed only prior to harvest. In perennial crops, all pipe work, save for sprinkler risers, may be buried.

These systems can be adapted to fields of any size and shape. Operating pressures and sprinkler heads can be selected according to crop water requirements and soil characteristics. Labour is required to open and close control valves unless full automation is provided. The major disadvantage of these systems is their high capital cost.

Continuous Move Sprinkler Systems

In continuous move systems the watering unit travels along a straight or circular path, irrigating as it goes.

Travelling Sprinklers

Travelling sprinklers may be sub divided into two types, those using flexible supply hose to reel in the sprinkler (most European models), and those using a wire cable and winch to pull a small wheeled carriage on which the sprinkler (rain-gun) is mounted. The winched machines are supplied by lightweight layflat hose whereas European models use stiff-walled hose.

Operating pressure is between 450 and 700 kPa at the gun, but friction losses in the hose add some 150 - 250 kPa at the hydrant. Travelling sprinklers will wet a strip up to 120 m wide and up to 400 m long, determined by the length of hose. Correct overlap of consecutive strips is essential to achieve uniform application. Under-irrigation occurs at field edges where there is no overlap. The high operating pressure and consequent high operating cost mean that travelling sprinklers are used mainly for supplemental irrigation. They require skilled operators and advanced maintenance facilities to keep them operational.

Centre Pivots

A centre pivot comprises a single galvanised steel lateral supported some 3 m above the ground on a series of wheeled A-frames. Hydraulic or electric drive units, mounted on the A-frames rotate the whole structure around a fixed, central pivot point. Water and power are supplied through the pivot point. A typical unit will have a total lateral length of 400 m, irrigating a circle of 50 ha within a 64 ha square. However, machines can range in length from 70 m to 800 m, irrigating areas from as little as 1.5 up to 200 ha.

The lateral supplies water to impact sprinklers, spinners or spray nozzles, the sizing and spacing of which along the lateral must take account of the differential speed of the lateral between inner and outer sections. To achieve uniform depth of application the rate of application varies along the length of the lateral. High application rates at the outer end can lead to surface ponding and run-off.

Centre pivots are suited to large, flat fields of uniform soil texture, free of any overhead obstruction. Very little labour is required for routine operation. Design, installation and maintenance require highly skilled staff and well-resourced workshops.

Low-Energy Precision-Application Technologies (LEPA) - These systems use drop hoses, attached to the main lateral on a centre pivot or linear move machine. A small orifice nozzle approximately 0.2 - 0.45 m above ground level irrigates below the crop canopy. Crops are planted so that hoses and nozzles hang between rows. Operating pressure at the nozzle is about 40 kPa. The spread of water from the nozzle is very limited, and localised application rates are high. Special tillage practices must be employed to form micro-basins or tied furrows to store water and prevent surface run-off. Growers may have problems maintaining adequate storage on the soil surface throughout the growing season. The system uses water more efficiently than conventional pivots fitted with sprinklers, as less water is lost to evaporation. Operating costs are also lower as a consequence of lower operating pressures. See: Lyle and Bordovsky (1983) and Hoffman and Martin (1993).

Linear Move Machines

Linear move machines use drive motors and alignment systems similar to centre pivots but the water supply is taken from a moving, rather than a fixed, point. Water is normally pumped from an open channel running in the centre of the field or along one boundary. Alternatively, it is supplied from fixed hydrants via flexible hose. The principal advantage of linear move machines over centre pivots is that the full field area is irrigated. However, the control and guidance systems of linear move machines are more complex. The machines can operate only in precisely rectangular fields with very little slope. The investment cost per irrigated ha is high but larger machines reduce the investment cost somewhat. Laterals are commonly 400 m or more in length.

Micro-Irrigation

Micro-irrigation applies water to only a fraction of the field surface, with water delivered through a network of pipes on to the soil directly or via small spray or bubbler (controlled orifice) outlets. Technologies are classified according to the type of flow control device used on the lateral. Almost all micro-irrigation systems are solid set, that is, sufficient mainline, manifold and lateral piping is installed to cover the entire crop area without movement of equipment. They require relatively little labour for routine operation once installed. The labour input is dependent on the degree of automation built into system control and the likelihood of damage and subsequent need for minor repairs to above-ground components.

Pre- and post-season labour demand varies greatly according to the type of system used and the degree of mechanisation. The annual cost related to pre- and post-season operations will vary between annual and perennial crops.

Where included in Table 1, the per hectare equipment costs are approximate. Cost is determined by crop type and row spacing. Figures are based on US data for large fields; unit costs for smallholders may vary considerably because fields are smaller and locally manufactured products may be available.

Water filtration is always recommended where drip emitters or line source laterals are used, to prevent blocking of the emitters, but sprayers or bubblers may in some circumstances be operated without filtration. The initial water quality and the type of emitter determine the degree of filtration. Chemical water treatment may also be required to prevent build-up of slimes or chemical precipitates.

Drip Emitters

Different types of emitter are used to control the flow of water from outlets on a lateral hose on the soil surface or buried in the crop root zone. The emitter may be factory-installed within the hose (in-line) or attached by a barb to the outside (side fitted/inserted). The lateral hose is typically flexible, thin-walled, polyethylene pipe with nominal inside diameters between 12 and 32 mm, 16 mm being the most common.

Most emitters, irrespective of type, are calibrated to operate at 100 kPa. The sensitivity of the discharge to changes in pressure is dependent on the type of emitter and the degree of compensation built into the design. Pressure compensation is normally effective over the range 100 - 300 kPa. At lower pressures discharge will be below the rated figure. Emitter design discharges normally lie in the range 2 - 8 l/hr.

Emitters are classified here according to the method used to dissipate pressure, which influences the form of the pressure/discharge relationship. This relationship has the general form:

$$q = Kd\, H^x$$

q	=	flow rate (l/hr)
Kd	=	discharge coefficient
H	=	design working pressure
x	=	discharge exponent

The lower the value of the discharge exponent the less the discharge will vary with changes in pressure. Keller and Bliesner (1990) provide a good overview of various types of emitter and selection criteria. The price of individual, barbed emitters varies widely according to type but will lie in the range $US 0.15 - 0.35 each. Polyethylene lateral hose may cost approximately $US 0.2 /m.

Long Flow Path

The earliest form of emitter used microtube or 'spaghetti' tubing - small bore polyethylene pipe with an internal diameter between 0.5 and 1.5 mm - pushed into a small hole in the lateral wall. The length of the microtube at different points is adjusted to achieve constant discharge as pressure changes along the length of a lateral. Much labour is involved in correctly sizing and installing microtubes. Tubes can easily become disconnected from the lateral. Long flow path emitters are slightly less likely than other emitters to become blocked, as the minimum flow path dimension (MFPD) for a given discharge is greater.

Moulded, long flow path, emitters also dissipate pressure along a long narrow conduit but the microtube is replaced by a screw thread, flat spiral or a moulded labyrinth. The discharge exponent of these devices lies between 0.5 and 0.8. Labyrinth (tortuous path) types lie in the lower part of this range, i.e they are less sensitive than spiral types to fluctuations in pressure.

Orifice Emitters

These are the simplest and cheapest type of emitter where discharge is controlled by the diameter of a small orifice. To supply low discharges of around 2 - 4 l/hr the orifice will have a diameter of 0.1 mm, a factor of 10 times smaller than in a labyrinth emitter operating at the same pressure. Relatively expensive manufacturing processes are needed to manufacture these very small orifices accurately and the orifice is very prone to blocking.

Many short flow-path orifice emitters now incorporate pressure-compensating mechanisms using flexible elastomeric disks which reduce the orifice diameter as pressure increases. By running the system at low pressure, with larger orifices, some degree of flushing is achieved.

A low-cost orifice drip system is described by Polak *et al* (undated). It is aimed at farmers cultivating 0.2 ha or less and is under trial in southern India and Nepal. Conventional drip systems are expensive because of the high number of laterals and individual emitters per unit area and the filtration system. The low-cost system uses moveable laterals to reduce cost. Each lateral is moved between 10 crop rows. Emitters are replaced by small (0.7 mm) perforations in the lateral, made with a heated needle. The perforation is covered by an outer sleeve, made from a 60 mm length of the same pipe material split lengthwise and slipped over the lateral. Discharge from the emitters is about 6 l/hr at 20 kPa and 10.5 l/hr at 40 kPa. Tests showed variations of up to 18% between emitter points. Simple mesh filtration is achieved using nylon cloth on the inlet to a 20 litre jerrycan. The equipment costs about $US 250 / ha.

Samani *et al* (1991) describe very similar equipment promoted in north east Brazil. Laterals are of 15 mm flexible PVC with emitter perforations of 1.2 mm. These are covered with a baffle in the same way as described by Polak *et al*. Emitter discharge is reported to be between 3 - 6 l/hr, at operating pressures of between 34 and 100 kPa, but emission uniformity was low.

Vortex Emitters

This emitter combines an orifice with a small vortex chamber where water spins and then passes out via a second chamber. The advantage over a simple orifice is that the minimum flow path dimension can be 1.7 times greater for a given pressure and discharge, thereby reducing the risk of the emitter clogging.

Line Source Tubing

Line source tubing irrigates a continuous strip along the length of the line. The width of the strip is determined by the soil texture and the discharge. Line source tubing normally operates at 30 - 70 kPa, a lower pressure than for point emitters. Modern line source "drip tape" is generally replacing external, manually placed barbed emitters for irrigating agricultural crops.

Single Chamber

The simplest type of line source comprises thin-walled pipe with small perforations at short intervals, 0.6 m or less, along its length. The disadvantage is that pressure varies greatly along the length of the lateral. Maximum lateral length is restricted to about 60 m to maintain adequate uniformity of distribution.

Double Chamber

Double chamber systems were developed to overcome the limitations imposed by pressure variations along single chamber line source laterals. An inner-chamber with widely spaced holes supplies water into an outer pipe with orifices spaced at intervals of 0.15 to 0.6 m. The inner pipe carries water at a relatively higher pressure whilst the pressure in the outer chamber is much lower. For each outlet in the inner pipe there are between 4 and 10 outlets in the outer pipe, spreading the discharge over a greater length of lateral.

Drip Tape

Double and single chamber pipes have largely been superseded by drip lines or drip tapes marketed under a range of brand names. Turbulent flow-path emitters are factory-inserted into the tape at intervals of 0.1 m up to 1.0 m. The design and minimum flow path dimension of emitters vary between manufacturers. Where laterals are manufactured for export and transport costs are high, some manufacturers favour very small emitters, which permit the lateral pipe to be rolled flat. This permits high-density packing in a container, but to achieve adequate flow control the emitter relies on a very small orifice with a consequent greater risk of clogging. The 'Drip-in' emitter is a larger component which takes up the full diameter of the lateral and prevents the pipe from being rolled flat. It uses a long labyrinth flow path and its minimum flow path for a given discharge is greater.

Drip tapes are supplied in a range of wall thicknesses from 0.15 mm to 0.38 mm. The discharge of each emitter lies in the range 1 to 2.6 l/hr and operating pressure is normally in the range 65 - 110 kPa. Tapes with pressure-compensating emitters are available, making it possible to operate long laterals at a higher inlet pressure and achieve high application uniformity. With careful handling, tape can have a working life of 10 years or more when used in annual cropping cycle.

Porous Wall Pipe

This type of line source lateral is not widely used due to clogging of the finer pores. Accurate control of discharge is not possible because pore size varies. Porous pipe has been manufactured from ABS (Acrylonitrile Butadiene Styrene), Polyethylene and PVC.

Sprayers

The general term 'sprayers' includes a range of products which produce a wetted circle with a diameter between 2 and 10 m dependent on operating pressure, rated discharge and elevation. Some types of sprayer may be mounted directly on to a lateral or on a stake 0.25 - 0.35 m high. Operating pressures vary between 50 and 500 kPa but typically lie in the range 100 - 300 kPa. Discharge varies from 20 to 250 l/hr.

Although primarily used for orchard crops - citrus, avocado, banana, etc - sprayers are also used to irrigate small areas of high-value fruit and vegetable crops. An individual sprayer, including the mounting stake, lead and coupling will cost between $US 1.15 and $US 2. Spray systems can often be operated without filtration or with a simple mesh or disk filter. Expensive sand filtration is not required.

Bubblers

Bubbler systems are used for the irrigation of orchard crops. A permanent, buried main and lateral network is installed. Each lateral serves one or two rows through individual bubbler outlets at the base of each tree. Bubbler systems provide a relatively high discharge - between 150 and 250 l/hr - at a point. Bubbler assemblies do not rely on very small orifices to control flow and therefore filtration is not required.

Unlike other micro-irrigation techniques which apply water little and often, bubbler systems apply a greater depth of water less frequently, storing larger volumes of water in the soil profile. Water is not sprayed or sprinkled over an extended area but flows as a low-pressure stream from the bubbler head. To prevent uncontrolled run-off and ensure a larger wetted area, water is contained within a shallow basin around each tree and allowed to infiltrate into the soil profile. Because larger volumes of water are applied at less frequent intervals than with other micro-irrigation systems, there is less risk of serious crop damage occurring, should the system temporarily fail, but conversely some authorities criticise bubblers for allowing inexperienced users to over-irrigate easily.

Pressurised Bubblers

Commercial bubbler systems operate at relatively high pressure, (100 -120 kPa). Pressure is dissipated in the bubbler head, which can often be adjusted in the field to vary the flow rate at each outlet.

Low Head Bubblers

Low-pressure bubbler systems are not marketed commercially. The discharge is not controlled by a device but by selecting the diameter, length and elevation of each discharge pipe. (See, Reynolds *et al*, (1995) and Hull, (1981)). In practice the design, installation and subsequent maintenance of these low head systems is labour intensive and requires a thorough understanding of the hydraulics of head loss in pipes.

Piped Distribution Systems for Surface Irrigation

Piped distribution systems potentially improve upon conveyance efficiencies of traditional open channel networks. Water is distributed to field hydrants or outlet boxes. From here it is conveyed to the field in open channels or by portable pipes or hoses such as layflat hose. Final delivery may be through gated pipes but water is still applied using surface methods. Van Bentum and Smout (1994) describe a number of different types of pipe distribution system based on the following criteria:

- Design operating pressure
    Low pressure - maximum design pressure < 100 kPa
    Medium pressure - Maximum design pressure between 100 - 200 kPa
    High pressure - Maximum design pressure > 200 kPa
- Pressure control
    Closed pipe network - no mechanism for dissipating excess head
    Semi-closed network - float valves used to regulate head between sections
    Open network - overflow standpipes serve as pressure break points between sections
- Scheme layout
    Looped system
    Branched system
- Origin of driving head
    Gravity
    Pumped

Of these different types the most widespread are low and medium pressure, closed networks with a branched layout, supplied from a pumped source.

When determining the design pressure rating of the system the designer must consider the maximum static pressure within any part of the system due to topography and the required pressure at any outlet which may be influenced by the possible future adoption of pressurised field application systems.

Piped distribution systems may be used by a single farmer to serve a single holding of 4 ha upwards, or may serve a number of farmers sharing a single outlet in turn. Tertiary-level networks serving many farmers over a command area of 100 ha or more have also been implemented. Cost per irrigated ha is very variable depending on the operating pressure of the network and local material costs. Capital costs lie in the range $US 800 – 2,500 /ha, depending on location.

Table 1   Summary of the Advantages and Disadvantages of Different Systems for Smallholder Farmers

System Type	Advantages	Disadvantages	Approximate Purchase Cost $US / ha	Approximate Labour Requirement man-hr /ha/irrigation
Hand-move Lateral	Easy installation; low purchase price; can be used in small and irregular fields; applicable to many types of crop.	High labour requirement; unpleasant work.	675 – 1,000 Low	1.73 High
Hopalong Systems	Easy installation; low purchase price; can be used in small and irregular fields; applicable to many types of crop. Reduced labour demand compared with conventional hand-moved laterals; less time lost in moves between sets.	High labour requirement and operating cost; unpleasant work.	Low	High
Drag Hoses	Easy installation; low purchase price; can be used in small and irregular fields; Applicable to many types of crop. Reduced labour demand compared with conventional hand-moved laterals. Can establish permanent/semi-permanent laterals and only move sprinklers.	High labour requirement; unpleasant work; flexible hoses can be damaged.	800 Low	2.0 High
Gravity Fed Artisan Sprinklers	Easy installation; low purchase price; easy to use and maintain; can be used in small and irregular fields; Applicable to many types of crop; different designs available for differing pressures and crop types.	Water use performance is inferior to proprietary products on pumped systems.	Low	High
Perforated Pipe or Sprayline	Easy installation; low purchase price; can be used in small and irregular fields; Applicable to many types of crop; small droplets present less risk of soil damage.	High labour requirement; oscillating mechanism prone to damage.	Medium	High
End Tow Lateral	Reduced manual labour and 'drudgery'.	Not suited to very small or irregular shaped fields; requires skilled operators to avoid damage when towing; crop area lost to turning areas.	1,100 Low/medium	0.62 Medium
Side Roll Lateral	Self-propelled unit; no land lost to turning areas.	Only suited to low-growing crops. Most efficient in large, unobstructed rectangular fields. Not appropriate for small irregular land holdings.	1,100 Low/medium	0.86 Medium
Side Move Lateral	More manoeuvrable than side-roll systems. Higher elevation allows irrigation of tall crops	Designed for use in large, rectangular fields. Complex drive mechanisms require high levels of maintenance. Not appropriate for smallholders.	1,300 Low/medium	0.62 Medium
Static Gun Sprinklers	Low cost; simple technology.	High labour requirement; high operating pressure; large droplets can cause soil damage. Not appropriate for smallholders.	800 Low	1.1 High
Boom Sprinklers	No clear advantages.	High labour requirement. Low application uniformity, especially on small fields of irregular shape. Difficult to manoeuvre between positions. Not appropriate for smallholders.	1,100 Low/medium	1.35 High
Fixed or Solid Set Sprinklers	Low labour requirement. Simple equipment with long economic life and low maintenance needs. Can be used in fields of irregular size or shape.	High capital outlay.	3,500 High	0.15 Low

Method	Advantages	Disadvantages/Constraints	Capital cost ($/ha)	
Travelling Sprinklers	Combines relatively low capital outlay and labour requirement. Little fixed in-field equipment.	High operating pressure; large droplets can cause soil damage. Skilled operators and advanced maintenance facilities required. Not appropriate for smallholders.	1,200 Medium	0.62 Medium
Centre Pivot	Very low labour requirement and relatively low capital cost. LEPA systems achieve high application efficiencies and lower operating costs.	Design, installation and maintenance require highly skilled staff and well-resourced workshops. Suited to large, flat fields of uniform soil texture, free of obstructions. Requires a reliable power supply. Must operate on a large scale to reduce investment cost /ha. Not appropriate for smallholders.	1,100 Low/medium	0.05 Low
Linear Move	No missed corner segments. Very low labour requirements.	Complex control and guidance systems require skilled operators and well-resourced workshops for maintenance. Can operate only in precisely rectangular fields with very little slope. Must operate on a large scale to reduce investment cost /ha. Not appropriate for smallholders.	1,500 – 2,000 Medium to high	0.1 Low
Drip Emitters	Very low labour requirement; high application efficiencies possible. Can be used in small, irregular fields and varying topography.	High cost. Requires skilled operation and maintenance to schedule irrigation and maintain filters.	2,500 – 5,000 High Cost / ha is highly dependent on row spacing.	0.05 Low
Line Source Emitters	Lower cost than point source emitters. Avoids need for manual insertion of emitters and easier to handle in the field.	High purchase cost. Requires skilled operation and maintenance to schedule irrigation and maintain filters. Emitters cannot be removed for manual cleaning or replacement. Single and double chamber laterals (bi-wall) are restricted in lateral length but drip tape overcomes this problem.	Re-usable 3,000 – 3,500 High Disposable 1,800 –3,000 High Cost / ha is highly dependant on row spacing.	0.05 Low
Sprayers	Minimal requirement for filtration. Can be used in orchards and for horticultural crops. Very low labour requirement.	Crop foliage may be damaged by wetting. Higher evaporative losses than with drip. Cannot be used with plastic mulches.	2,500 – 3,200 High	0.05 Low
Pressurised Bubblers	No requirement for filtration. Suited to small and irregular field shapes.	Only applicable to tree crops.	2,500 – 4,000 High	0.05 Low
Piped Distribution	Low technology with minimal maintenance requirements. Suited to a wide range of crop types	Requires careful design and high construction quality for effective operation. Limited water savings. Systems requiring co-operative water management may not be effective.	800 – 2,500 High	High

3 Matching Technologies to the Needs of Smallholders

This chapter analyses the technical characteristics of the irrigation systems described in Chapter 2 relating them to the conditions and requirements of smallholders. Wider issues, not directly linked to the nature of the equipment itself, are brought out in Chapter 4 and summarised in Chapter 6.

The Context

The development of modern irrigation technologies has been driven by the need to reduce labour and other operating costs and improve water use efficiencies. Most systems are designed to meet the needs of medium and large-scale, high input, commercial agricultural enterprises. The benefits of modern technologies for such users include reduced costs, improved yields, improved water use efficiencies and others (see Table 2). It is important to consider whether these benefits can be realised by smallholders in less developed countries. Hillel (1989) warns of a gap between high technology systems and the needs of small-scale farming in arid regions of the developing world where the benefits of drip (and other technologies) could be most marked. He suggests that researchers and manufacturers are fascinated by high technology, developing ever more specialised and intricate hardware and states that:

"In the non-industrial countries, the important attributes are, low cost, simplicity of design and operation, reliability, longevity, few manufactured parts that must be imported, easy maintenance and low energy requirements."

Hillel also suggests that "Labour economy is less important" although this is dependent on local conditions and the availability and cost of hired labour.

Keller (1990) maintains that the principal benefits offered by modern irrigation systems are higher water use efficiencies, through reduced conveyance losses and improved field application, coupled with greater control over the timing and depth of applications. By adopting a modern irrigation method, the farmer can achieve higher productivity per unit of water and land.

Under traditional irrigation methods, the productivity of water is limited by a farmer's willingness to invest labour and management skills in accurate land levelling and field preparation and in 'coaxing' water to spread evenly over the field surface. The purchase of a modern, pressurised irrigation system, of whatever type, trades money for labour and skill (Keller, 1990). In many situations the opportunity cost of money for the smallholder is very high whilst that of labour and traditional skills is low. Farmers will make the investment in modern technologies only when the financial return is clear and relatively assured.

Where smallholders are profit maximisers they will aim to minimise production costs and maximise returns to inputs by increasing the quantity or quality of production. Given these objectives, modern irrigation technology is likely to be attractive where it can reduce high production costs, that is, where the cost of water is high, and/or where higher yields can be marketed at a profit.

Subsistence farmers are mainly driven by the need to minimise risk and assure a food supply rather than by market forces and a wish to maximise profit. In such cases, new irrigation technology might only be considered where it offers more secure production of basic foods and reduced risk of crop failure with minimal expenditure. This is seldom seen in the field. Rather, new irrigation technologies are adopted to support cash crop production as part of a package that moves farmers from subsistence to commercial production.

Table 2   Potential Technical Advantages of Modern Irrigation Systems for the Smallholder

Sprinkler Technologies
- Improved conveyance and application efficiencies on coarse textured and shallow soils
- Low discharges may be used
- Applicable on undulating and steep terrain without need for land forming (Gravity head may be used to pressurise the system)
- Reduced labour requirements

Micro-irrigation Technologies
- Maintain favourable soil moisture conditions on poor soils - gravels, coarse sands, clays
- Applicable on undulating and steep terrain without need for land forming
- Drip and bubbler systems unaffected by wind (compared with sprinkler)
- Permit use of poor quality water
- Permit accurate application of fertilisers
- Avoid leaf scorch and reduce risk of foliar fungal disease (compared with sprinkler)
- Localised soil wetting reduces evaporative losses and weed growth between rows
- Operate at lower pressure than sprinklers, thereby saving energy

Piped Conveyance Technologies
- Improved conveyance efficiencies (compared with open field channels)
- Absence of field channels provides more land for crop production and easier cultivation

Water savings

Conveyance efficiency:	Open field channels	70%
	Piped distribution	80 - 85%
Field application efficiency:	Surface methods	50 - 60%
	Sprinkler	70%
	Micro-irrigation	80 - 90%
Overall efficiency:	Surface methods	38%
	Sprinkler	57%
	Micro-irrigation	70%

Technical Factors Influencing Uptake of New Technology

In a review of modern irrigation technologies in developing countries, Keller (1990) suggests a number of technical factors, relating to operation and maintenance, that determine whether a smallholder will take up a system successfully. For each factor several categories are identified and "scored". The scores allocated to each of the categories are indicated in the text – higher values reflect greater suitability for smallholders. Some element of weighting is built-in against technologies that are non-divisible by allocating a score of zero rather than one to this category. Scores for each factor are summed and the technologies ranked according to the total score (Table 4). The assignment of categories to technologies is based on subjective assessment and the resulting ranking should be used only as an approximate guide to the relative suitability of different technologies for smallholders. The ranking does not take account of the operating or capital cost of equipment but indicative values of capital cost per ha are given in the table, based on large-scale installations in the United States.

Divisibility
The suitability of the technology for use on small and irregular shaped land plots of 0.2 to 5 ha.

- Well-suited for use on any area and shape of plot. Implies that supply, distribution and field application equipment can be operated by an individual farmer: Total [3]

- Only applied with difficulty and/or high expenditure to small plots. Normally implies some group co-operation to control water supply or distribution between users: Partial [2]

- Technologies only suited for use on large and regular-shaped plots: None [0]

Maintenance
Indicates the complexity of the maintenance task and possible requirement for specialist technicians to carry out maintenance.

- Only basic skills, easily acquired by a 'traditional farmer', required to maintain the equipment: Basic [4]

- Can be maintained by the farmer but requires skills associated with more entrepreneurial farmers growing high-value crops: Grower [3]

- Some specialist skills or equipment required: Shop [2]

- Specialist technicians with workshop facilities and equipment are needed for maintenance: Agency [1]

Risk
Indicates the risk of serious yield reduction or crop loss as a consequence of equipment failure.

- Risk of component failure is slight and problems can be easily rectified. Soil moisture storage normally provides an adequate buffer against a brief shutdown: Low [3]

- Failure of a component would only jeopardise the supply to a single outlet: Medium [2]

- Failure of a single component can result in complete shutdown of the system. (Applies to drip systems requiring micro-filtration and all continuous move systems): High [1]

Operational Skill
Indicates the level of training and understanding required of the operator to achieve good water application efficiencies and avoid damage to the equipment.

- Few skills, easily acquired during a single season's operation, are required: Simple [3]

- Considerable skill and care are required to operate the equipment effectively without damage: Medium [2]

- Needs good understanding of the system design and operating principles and/or extended field experience to achieve good application efficiency: Master [1]

- The user must acquire complex technical skills to operate and service the equipment: Complex [1]

Durability
Indicates the likelihood of equipment breakdown during normal operation and susceptibility to damage as a result of improper handling or operation.

- Systems with few or no moving parts, other than in the pump. Unlikely to break down: Robust [4]

- Systems not likely to suffer breakdown or damage through improper handling but none the less requiring a minimum of spares for immediate repairs and periodic servicing: Durable [3]

- Systems that require careful operation and extensive workshop and spares backup to remain operational: Vulnerable [2]

- Systems highly prone to breakdown if subjected to inadequate maintenance or incorrect operation: Fragile [1]

Table 3 shows the results of applying these criteria to the technologies described in Chapter 2. The results of ranking the technologies on the basis of the scoring system are shown in Table 4.

Table 3   **Factors Influencing the Appropriateness of Different Irrigation Systems for Smallholders (After Keller and Bliesner, 1990).**

System Type	Divisibility	Maintenance	Risk	Operational Skill	Durability
SPRINKLER **Periodic move**					
Hand-move	Total (3)	Shop (2)	Med (2)	Medium (2)	Durable (3)
Drag hose	Total (3)	Basic (4)	Med (2)	Simple (3)	Durable (3)
Low tech. Sprinkler	Total (3)	Basic (4)	Low (3)	Simple (3)	Durable (3)
Perforated pipe	Total (3)	Shop (2)	High (1)	Simple (3)	Vulnerable (2)
End-tow	Partial (2)	Shop (2)	Med (2)	Medium (2)	Durable (3)
Side-roll	None (0)	Shop (2)	High (1)	Medium (2)	Vulnerable (2)
Side move	None (0)	Agency (1)	High (1)	Complex (1)	Fragile (1)
Static gun	Partial (2)	Shop (2)	Med (2)	Master (1)	Durable (3)
Boom sprinkler	None (0)	Shop (2)	High (1)	Master (1)	Vulnerable (2)
Solid Set					
Portable	Total (3)	Shop (2)	Med (2)	Simple (3)	Durable (3)
Permanent	Total (3)	Grower (3)	Med (2)	Simple (3)	Durable (3)
Travelling gun	Partial (2)	Agency (1)	High (1)	Master (1)	Vulnerable (2)
Centre pivot	None (0)	Agency (1)	High (1)	Complex (1)	Vulnerable (2)
Linear move	None (0)	Agency (1)	High (1)	Complex (1)	Vulnerable (2)
MICRO-IRRIGATION					
Drip emitters	Total (3)	Grower (3)	High (1)	Complex (1)	Fragile (1)
Line source:					
Reusable	Total (3)	Grower (3)	High (1)	Complex (1)	Fragile (1)
Disposable	Total (3)	Grower (3)	High (1)	Complex (1)	Fragile (1)
Sprayers	Total (3)	Grower (3)	Med (2)	Complex (1)	Durable (3)
Bubbler					
Pressurised	Total (3)	Grower (3)	Low (3)	Simple (3)	Robust (4)
Low pressure	Total (3)	Grower (3)	Low(3)	Master (1)	Vulnerable (2)
PIPED CONVEYANCE					
Piped distribution	Total (3)	Grower (3)	Low (3)	Simple (3)	Robust (4)

Table 4   **Ranking of System Types Reflecting Suitability for Smallholders**

System Type	Score[1]	Crop Types	Initial cost $ US/ha[2]
Piped distribution	16	All types	800
Low tech. Gravity fed sprinkler	16	All types	N/a
Pressurised bubbler	16	Orchard	3,000
Drag hose, sprinkler	15	All types	675
Permanent solid set sprinkler	14	Orchards; soft fruit	3,500
Portable solid set sprinkler	13	All types	3,250
Hand-move sprinkler laterals	12	All types	675
Micro-irrigation sprayers	12	Orchard, soft fruit and vegetables	3,500
Low pressure bubbler	12	Orchard	3,000
Sprinkler, perforated pipe	11	Soft fruit and veg.	800
Sprinkler, end-tow lateral	11	Cereal and row crops	950
Sprinkler, static rain gun	10	Cereal and row crops	N/a
Drip emitters	9	Wide row fruit/veg; Orchard	3,500
Line source reusable	9	Wide row fruit/veg	5,000
Line source disposable	9	Wide row fruit/veg	3,000
Sprinkler side-roll	7	Short cereals and row crops	1,100
Travelling rain gun	7	Cereal and row crops	1,200
Boom sprinkler	6	Cereal and row crops	N/a
Centre pivot	5	Cereal and row crops	1,500
Linear move	5	Cereal and row crops	1,300
Sprinkler side-move	4	Cereal and row crops	1,350

Notes:
1.   Maximum score = 17
         Minimum score = 4

2.   After Keller (1990). Costs are based on US experience and include mainlines and pumping plant with systems installed on large fields.

Systems that are technically more appropriate for use by smallholders include:

- Piped distribution systems for surface irrigation
- Low technology sprinklers
- Pressurised bubbler (orchard crops only)
- Drag hose sprinklers

These are relatively low technology systems, easily adapted to small plots, easily maintained and requiring limited skills of the operator. With the exception of bubbler systems they have low capital costs, which must be traded-off against higher labour requirements to move equipment manually around the irrigated area. The much higher capital cost of a pressurised bubbler system reflects the fact that this is a set system.

Labour requirement and energy costs are not included in this ranking procedure. Labour requirements will vary according to field layout, topography and crop type as these influence equipment density, the number of independent blocks that must be controlled and the time between set changes. In general, sprinkler and piped conveyance systems have a higher demand for labour than micro-irrigation systems. The sprinkler systems with the least labour requirements are either too costly or too complex for smallholders. Energy costs are highly variable, depending on the energy source – electricity, petrol, kerosene or diesel – and location. Systems with lower operating pressures incur lower energy costs. Piped distribution networks and low technology sprinklers are often installed where the gravity head is sufficient for their operation. Pressurised bubbler systems operate at 100 – 120 kPa. This is at the low end of the range of operating pressures found in commercial systems. Drag hose sprinklers using conventional sprinkler heads require between 200 – 400 kPa at the sprinkler nozzle. Pumping costs in these systems can represent a significant variable production cost limiting their use to higher value crops. Where farmers pay for energy according to the volume of water pumped this can provide a powerful incentive for the adoption of water conservation measures.

Bubbler systems for orchard crops are the only form of micro-irrigation technology that ranks highly on the basis of technical suitability. Of the other forms of micro-irrigation, mini-sprinklers or sprayers rank higher than drip and line source systems. In these systems there is less risk of widespread system failure as the consequence of a key component failing, and the equipment is less prone to damage through poor operating practice.

Large irrigation machines such as centre pivots, linear move laterals and continuous move rain-guns are at the bottom of the list. Their mechanical complexity and non-divisibility for small land holdings are normally regarded as making them unsuitable as a method of irrigation for smallholders. Keller (1988) and Manig (1995) describe the possibility of using large-scale modern irrigation hardware, such as centre pivots, under the administration of a state agency, to provide "controlled rain" to many smallholder plots. However, there are few examples of this development option being implemented, although examples can be found in South Africa where small centre pivots (40 ha) exist, established and maintained by government agencies, and supplying irrigation to four farmers (Louw, 1996).

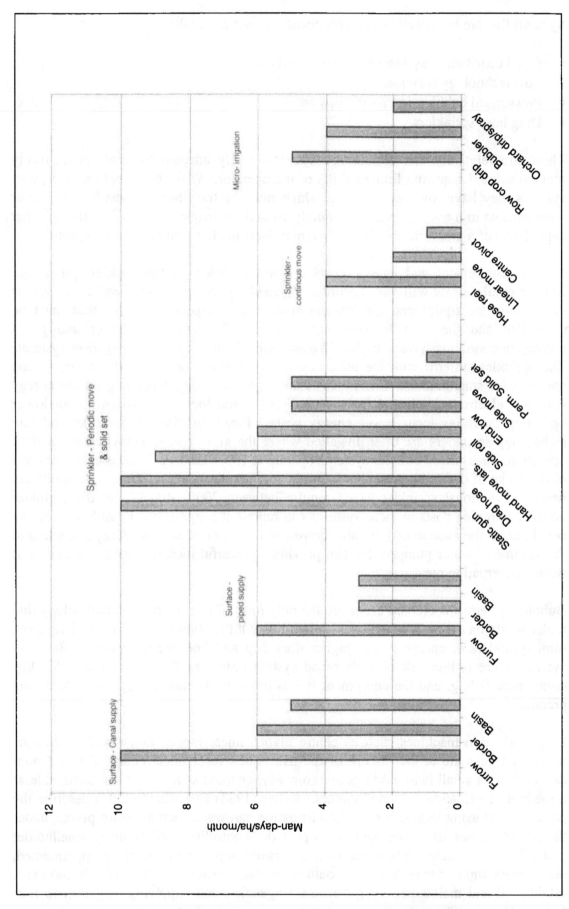

Figure 3   Labour Requirements of Different Irrigation Methods

4 The Uptake of Technologies – Experiences to Date

This chapter summarises the experience of smallholders with modern irrigation technologies in a range of economic and agro-ecological conditions. The term 'smallholder' is defined in Chapter 1. Essentially the term is used for all farmers concerned with agriculture as the main source of food and income for the household; large-scale, commercial farming is not considered.

A wide variety of production systems are considered, from high technology vegetable production for export in the arid lands of Jordan and Israel, to low technology systems serving local markets in regions as diverse as semi-arid northern China, monsoonal India and semi-humid Guatemala.

Factors affecting the willingness and ability of smallholders to adopt and maintain modern irrigation technologies are identified from field experience and a comprehensive literature review. The review indicates that the factors having the greatest influence on the uptake of modern irrigation systems can be grouped under the following headings:

- Availability of water
- Form of exploitation - Individual or communal schemes
- Farming system
- Role of government and private sector agencies
- Marketing and finance

Table 6 summarises information relating to the use of modern irrigation by smallholders in eleven countries. More detailed information on individual countries is included in Appendix 2.

Availability of Water

A number of different terms relating to water scarcity are used in the literature, including water shortage, water scarcity and water stress. Winpenny (1997) defines shortage as an absolute measure defined in terms of renewable volume per head per year, whilst scarcity indicates an imbalance between supply and demand. Thus, countries with no obvious water shortage may face scarcity due to excessive consumption. Water stress describes the symptoms of shortage or scarcity and cannot itself be directly enumerated. Differences arise over measures of national water availability, resulting from the use of either global or internal resources, i.e including or excluding cross-border surface flows. Notwithstanding this uncertainty of definition, the following categories are used by a number of authorities, including the World Bank and FAO, to classify water shortage:

10,000 - 2,000 m^3/caput	Water management problems
2,000 - 1,000 m^3/caput	Water stress - large investments required
< 1,000 m^3/caput	Absolute water scarcity

(Van Tuijl, 1993)

Of 18 countries estimated to have water resources equivalent to less than 1,000 m^3/head/yr in 1994, 13 are in North and Eastern Africa (FAO, 1995a) and the remaining 5 in the Middle East (Van Tuijl 1993). It is predicted that Sudan and

Morocco will also fall into this category by the year 2000 (FAO, 1995a). The data in Table 5 indicate that there is a relationship between water availability and the adoption of advanced modern irrigation technologies. However, water scarcity is not the sole factor determining the extent of adoption. Many of the North and East African states referred to above face equal or more severe water shortage than the countries in Table 5 but do not show the same levels of adoption.

Table 5   Water Availability and the Adoption of Micro-Irrigation Technology

Country	Micro-irrigation as % of total irrigated area	Internal water availability m³/head/yr
Cyprus	71	1,250
Israel	49	431
Jordan	21	405
South Africa	13	1,105

For many nations a national figure of water availability can mask important regional variations. India is a clear example where the national average figure in 1981 was 1,353 m³/head/yr, but figures for individual States varied from 788 for Maharashtra to 3,450 for Punjab. China is a further example where regional variation is great, but spatial variation in supply can also be significant in smaller nations such as Kenya, Senegal and Ethiopia.

Israel and Jordan face extreme water scarcity, with only 400 - 450 m³/head/yr available. Cyprus faces less extreme problems with 1,250 m³/head available. In all three countries there has been major government investment in water conveyance and distribution infrastructure delivering pressurised supply at the farm outlet. Provision of this infrastructure, along with other factors including water pricing policies, effective agricultural extension and private sector investment in equipment supply has led to widespread adoption of modern irrigation technology. Modern irrigation of some form serves all of Israel's irrigated land. In Cyprus, sprinkler or micro-irrigation has been adopted on 96% of land served by public infrastructure and in Jordan, 68% of the irrigated area is under modern systems.

Egypt has a long history of surface irrigation and consequently has made large investments in gravity irrigation infrastructure. With water availability at 1,062 m³/head/yr, Egypt faces greater water scarcity than Cyprus, yet modern technology[2] is used on only 13% of the irrigated area. Where new infrastructure has been built in the New Lands, water is pumped up from the Delta and conveyed in open channels. In contrast with the public conveyance systems in Jordan and Cyprus, water is not supplied under pressure at the farm turnout. The Egyptian Government, unlike those of Cyprus, Jordan and Israel, does not charge farmers for water. This is one important distinction that contributes to the relatively low rate of adoption of modern technology.

---

[2] Whilst this percentage appears low by comparison with Cyprus, Jordan or Israel the area reportedly using modern technologies in Egypt is 1.3 times greater than the area in these three countries combined.

Table 6   Summary of Smallholders' Experience with Modern Irrigation Technologies – by Country

Country	Irrigation types and Areas[1] (% of total Irri. area)	Water supply		Farming system		Govt. and Private sector		Marketing & finance	
		Water Availability m³/head/yr	Water source	Average farm holding[11]	Crop	Level of Govt. support	Principal agents of promotion	Market for produce	Price subsidies & Incentives
Israel	Total modern: 193,000 ha[2] (100%) Micro: 104,000 ha Sprinkler: 89,000 ha	431	Mixed	2 – 4 ha (Regev et al, 1990)	Vegetables Orchards Field crops	Major commitment in infrastructure & policies	Govt. agencies	Favourable local and export	Early financial incentives for local manufacturers. Soft loans to farmers.
Cyprus	Total modern: N/a Micro: 25,000 ha[3] Sprinkler: N/a	1,250	Govt. schemes exploiting gravity head. Private wells	< 1 ha (Van Tuijl, 1993)	Vegetables Orchards Grapes	Major commitment, including infrastructure & land consolidation	Govt. agencies	Favourable local and export	Govt. funded infrastructure + subsidies to farmers buying equipment.
Jordan	Total modern : 43,600 ha[4] (68%) Micro: 38,300 ha Sprinkler: 5,300 ha	405	Govt. schemes exploiting gravity head. Private wells in highlands	3 – 5 ha (Hanbali et al, 1987)	Winter vegetables	Major commitment, including infrastructure & land consolidation	Private sector	Favourable export	Public funded infrastructure.
Egypt	Total modern: 416,000 ha[5] (13%) Micro: 104,000 ha Sprinkler 312,000 ha	1,062	Small pumps from canals and wells	3 ha	Winter vegetables Orchards Groundnuts Field crops	Canal infrastructure Cheap credit	Private sector	Favourable local and export	Soft loans for equipment purchase
India	Total modern: 131,800 ha (0.2%) Micro: 55,000 ha[6] Sprinkler: 76,800 ha[7]	1,353 (Avge) Highly variable between states.	Tube-wells – individual and co-operative ownership	Sprinklers 1 – 5 ha (Shelke et al, 1993) Drip 0.8 - 2 ha (Sivanappan, 1988)	Vegetables Orchards Cut flowers	Moderate. State subsidies up to 50% on equipment purchase	Private sector manu-facturers	Local	Subsidies and soft loans for equipment purchase.
China	Total modern: 3,170,000 ha (7%) Micro: 20,000 ha[8] Sprinkler: 650,000 ha[9] Low pressure Pipes: 2,500,000 ha[8]	2,420 (Avge) Highly variable between regions.	Tube-wells	State owned.	Vegetables Orchards Field crops	Minor. Policies for water saving in place	Govt. agencies	Local	Low interest loans

	Total modern		Water source	Plot size	Crops	Extent	Agency	Market	Credit
Pakistan	Total modern: N/a Sprinkler: Very small	1,467	Tube-wells	Sprinkler sets for plots from 0.2 – 20 ha	Field crops Orchards	Very Minor		N/a	N/a
Guatemala	Total modern: N/a Gravity Sprinkler: 2,000 ha[10] (2.5%)	11,959	Springs	0.2 – 1.4 ha (Lebaron, 1993)	Vegetables Flowers Maize Beans	Targeted project	Govt. agencies	Favour-able local	Soft loans (2% interest over 20 yrs)
Sri Lanka	Total modern: N/a Low Head Drip & Drip - very small areas	2,586	Shallow hand dug wells	Trial plots 1 ha and 2 ha	Vegetables	None	N/a	Local	N/a
Kenya	Total modern: 22,000 ha[5] (33 %) Micro: 1,000 ha Sprinkler: 21,000 ha	1,119	Rivers – pumped and gravity	1 ha	Vegetables	None	Private sector	Local and export	Advice in locating sources of credit
Zimbabwe	Total modern: 95,000 ha[5] (82%) Micro: 8,000 ha Sprinkler 87,000 ha	1,923	Wells Reservoirs Rivers	0.5 ha and below	Vegetables Groundnuts Some field crops	Major commitment. Design, implementation and extension.	Govt. agencies	Local and export	Major infrastructure given to farmer groups.

Notes:

1. Areas shown are totals under modern methods. A large proportion is found on large commercial farms and estates.
   Sources:
2. Nir (1995)
3. Van Tuijl (1993) and Field (1990)
4. Battikhi and Abu-Hammad (1994)
5. FAO (1995b)
6. Bucks (1993)
5. Sharma & Abrol (1993)
6. Kezong (1993)
7. Chen Dadiao (1988)
8. Lebaron (1993)
11. Farm holding size corresponds to average smallholder

India, Pakistan and China also have extensive public canal systems, serving the needs of surface irrigation. All three nations have average water availability figures that indicate impending water shortages - India 1,353 m^3/head/yr, Pakistan 2,088 m^3/head/yr and China 2,360 m^3/head/yr (United Nations ESCAP, 1995). Surface irrigation is still the norm in these nations and there is no evidence of planned integration of canal conveyance with modern field application methods. Where modern systems are used in these countries the water source is normally a private or co-operatively owned tube-well.

Form of Exploitation - Individual or Communal Schemes

Schemes where farmers are required to share water or field equipment below an outlet appear more difficult to sustain. Joint ownership of sprinkler equipment on the Doukkala II project in Morocco led to poor maintenance of the equipment (Van Tuijl, 1993). Hinton *et al* (1996) report on the difficulties faced in operating a piped distribution system in Egypt where a small number of group outlets replaced numerous traditional field watercourses. On large public schemes those that have been successful are those where water is delivered to individual farm turnouts under pressure. Pressure regulation and primary water filtration on these large schemes are the responsibility of the government agency rather than the farmer.

In Guatemala, mini-irrigation projects allow groups of hillside farmers to exploit small springs to irrigate land that would be unirrigable by surface methods. The PVC pipe distribution network provides each farmer within an established association with one or more farm off-takes. All members of the association share ownership and maintenance of the distribution network. The early systems were designed to operate 'on-demand' to avoid the need for farmer co-operation in scheduling irrigation, but later designs use smaller pipes, at lower unit costs. Farmers on these schemes must co-operate and take water in turns. Lebaron *et al* (1987) do not report how successful this introduction of more complex management practices has been.

In Zimbabwe the Government has promoted smallholder schemes using draghose irrigation for farmers with holdings of 0.5 ha on average. All farmers on a scheme share responsibility for the operation and maintenance costs of pumps and the piped distribution network but individuals have their own draghoses and sprinklers. Water is taken in turn between holdings situated next to the lateral.

In Kenya, individual farmers responded to water shortage at the tail of a surface system by purchasing locally manufactured butterfly sprinklers operated by gravity from the canal system. The sprinklers were cheap, robust and reliable. Uniformity of water application and efficiency were low by comparison with commercial equipment but still a distinct improvement over surface methods. Application efficiency was about 65%.

In India there is little evidence of smallholders collaborating to exploit a water supply using modern technology. In almost all cases an individual farmer draws water from a shallow or deep tube-well. An exception is the sprinkler irrigation co-operative described by Rao (1992) comprising 16 marginal farmers in Karnataka State. He reports that support of the co-operative is weak and concludes that further extension is essential to secure the widespread adoption of sprinkler systems by marginal farmers in the dry zone.

The Sub-Regional Workshop on Irrigation Technology Transfer in Harare, (FAO/IPTRID, 1997), laid heavy emphasis on the potential role of low cost treadle pumps in expanding smallholder irrigation in the region, citing widespread use of such pumps for irrigation in Bangladesh. This manual water abstraction technology can be applied only where the source is either surface water adjacent to the field plot or shallow groundwater (at 5m or less) underlying the plot. There is little evidence as yet that this low technology technique will be widely adopted by farmers in Africa.

There is no one type of water source - groundwater, river or reservoir - or form of exploitation - individual owner or communal/project development - that lends itself particularly to exploitation by pressurised technology.

Farming System

Field Size and Land Tenure

Micro-irrigation technologies, conventional sprinklers and draghose systems can be operated successfully on very small field plots. For example, hillside farmers in Guatemala irrigate plots as small as 0.2 ha, drawing water from a community-owned pipe system that may supply 5 to 30 ha (Lebaron, 1993). In Karnataka, India, Rao (1992) describes a farmers' co-operative including 16 farmers, each having a holding of less than 1ha, operating a conventional hand-move lateral sprinkler system. Although the technical sustainability of this scheme is uncertain, the small size of plots was not a constraining factor. Sivanappan (1988) reports the widespread use of drip technologies by farmers irrigating plots of 2 acres (0.8 ha) or less in India. In Zimbabwe farmers use draghose equipment on plots of 0.5 ha and less. There are therefore a number of modern technologies that can be successfully applied on small land holdings.

Security of land ownership is often a prerequisite for farmers to secure loans or grants for the purchase of irrigation equipment. In both Jordan and Cyprus, important land consolidation legislation was enacted to secure uniform plot sizes, prevent land fragmentation and provide security of tenure to farmers. Formal land registration was required of farmers forming associations to take advantage of the mini-irrigation project in Guatemala (Lebaron, 1987). In southern and eastern Africa, Rukuni (1997) argues in favour of recognising and strengthening traditional land tenure systems that were neglected and over-ruled by colonial and contemporary governments in favour of state ownership. Reform of the existing situation would transfer property rights to land and water back to communities who would oversee their allocation to individuals.

Farmers require security of tenure before investing in expensive irrigation equipment. Where government-funded projects are established, the rights and responsibilities of farmers on the project must be well understood and accepted by all partners.

Crop Type

In almost all cases identified, modern irrigation equipment is used to irrigate high value cash crops marketed off the farm. Systems are seldom used by smallholders to irrigate subsistence crops. Modern systems may be used by large-scale commercial-sector farmers to irrigate staple grain or other field crops.

It is notable that even where a technology permits the irrigation of basic grains, such as small portable rain-guns in Pakistan, the equipment is none the less used for production of high-value vegetable and fruit crops (Irrigation Systems Management Research Project, 1993).

Guatemala, China and to some extent Zimbabwe, provide the only examples of modern systems used by smallholders to irrigate field crops. Lebaron *et al* (1987) report that on a small number of the mini-irrigation projects in Guatemala using gravity head sprinklers, farmers sometimes grow a second crop of maize or beans - traditional staples - in the dry season. However, the major impact of the irrigation systems has been to develop dry season irrigation of non-traditional crops such as carrots, cauliflower, onions, chilli and strawberries, which are marketed in Guatemala City or other local centres.

In northern China, low pressure low-cost piped distribution systems are widely used to irrigate grain crops using conventional basin or furrow irrigation methods at the field level. In Zimbabwe, low technology low-cost buried clay pipes have been evaluated and promoted, but the technology lends itself to the irrigation of small areas of high-value vegetables rather than staple grains. On schemes using draghose irrigation in Zimbabwe farmers may grow maize but the greater emphasis is on potatoes, groundnuts, onion, cabbage, green beans, peas and other green leaf vegetables.

Rukuni (1997) states that "The marketing and trade of irrigated high-value crops offers the greatest opportunity for intensifying small-scale irrigation in East and Southern Africa."

Other Inputs

Pressurised irrigation systems were introduced to small, traditional farmers in parts of Jordan (Van Tuijl, 1993) and Israel (Keen, 1991) as part of larger agricultural development and extension programmes, alongside improved seed varieties, and increased use of fertiliser, herbicide and pesticide. Irrigation technology was thus part of a package of measures that promoted a shift from traditional agriculture to intensive production and marketing of winter vegetables. In Zimbabwe, where draghose sprinklers are used, farmers have access to good extension and use hybrid seeds, pesticides and inorganic fertilisers.

By contrast, in Guatemala, gravity-driven sprinklers were promoted in isolation from other agricultural extension initiatives. It was intended that the equipment should be used to grow non-traditional cash crops, but Lebaron *et al* (1987) report that supporting agronomic extension information was lacking from the project. However, farmers responded positively and moved to the production of non-traditional crops through a trial and error process over a period of time.

Farmers must achieve good early returns on investment. Such returns can be achieved only by a package of measures with which traditional farmers may not be familiar and there is therefore a need for advice from specialised irrigation agronomists.

The Role of Government and the Private Sector

In Israel, Jordan and Cyprus, government policy decisions to conserve scarce water resources resulted in major infrastructure projects, which in turn facilitated the adoption of pressurised irrigation by farmers. In Israel and Cyprus, effective public extension services well trained in irrigation engineering and irrigation agronomy, also played an important role in promoting modern technologies amongst smallholders, (Melamed, 1989; Van Tuijl, 1993). In Israel, financial support from government was available to assist early manufacturers of drip irrigation equipment.

In Egypt, the government prohibits surface irrigation on the sandy soils of the New Lands areas, in order to reduce water use and the dangers of waterlogging and salinity. Farmers are offered low interest loans for irrigation equipment, repayable over 20 years. However, in the absence of adequate technical extension advice, many traditional farmers have reverted to surface methods in the face of technical problems and poor returns. Agricultural graduates settled in the same area have persisted in the use of sprinkler systems but have faced technical difficulties due to poor design leading to incorrect equipment and layouts for the crops being grown. Specific problems include:

- Poorly installed pipe systems
- Pump sets not matched to the required pressure / discharge characteristics
- Poor understanding of crop water needs and lack of information about timing and depth of applications.

In Jordan, Van Tuijl (1993) reports that private sector companies imported and sold equipment and also provided a design service. The same companies also provided cheap credit to farmers when the state banks were still unconvinced of the merits of micro-irrigation systems.

The Indian government has sought to promote modern irrigation technologies primarily by subsidising purchase of equipment. The value of the subsidy varies between states and also depends on the farmer's total landholding and economic status. Despite subsidies of as much as 75% of total equipment cost, Saksena (1993) reports that the rate of adoption is very slow and uneven.

Drip technology has been adopted most readily in Maharashtra State. 66% of India's total drip irrigated area was in Maharashtra in 1993 (Singh *et al*, 1993). The concentration of drip irrigation in this state is due to a number of factors including water shortage - Maharashtra has the lowest water availability per head of all Indian states - relatively affluent farmers and good access to markets. Jain Irrigation Systems Ltd, located in Bombay, has played an important role in promoting drip technology. The company is one of the largest manufacturers of drip irrigation equipment in India, manufacturing equipment under licence from a major American company. Jain offers assistance in system design and installation, providing some maintenance and after-sales care to farmers wishing to install drip systems. The combination of state subsidies for equipment purchase - available in every state - together with village demonstration plots and effective technical support for design and installation offered by Jain and a large market in Bombay for high value products, has contributed to the concentration of drip technology in this state.

Despite the efforts of the government extension services and the better-resourced private companies, two surveys of smallholders' drip irrigation systems in Maharashtra, (Holsambre, 1995; Dalvi *et al* 1995) showed that installations were failing to provide the high water use efficiencies theoretically attainable from such systems. Major faults were:

- Mismatch of pumps with the pressure/discharge requirements of drip systems
- Inadequate filtration
- Leakage at joints due to poor installation
- Pressure variations owing to inadequate allowance for land slope.

Despite faults, farmers continue to operate the systems, recognising the benefits of savings in water and labour over surface irrigation. However, less than quarter of the sample reported improvements in yield.

In Guatemala, mini-irrigation projects amongst hill farmers were promoted under a joint project between USAID and the Guatemalan Government extension service. After some early pilot installations, the spread of the technology depended on village groups approaching the extension service and requesting assistance in system design and installation. Loans for equipment are provided by the state agricultural bank at very low rates of interest, repayable over 20 years. About 250 systems were established over an eleven-year period up to 1989 serving a total of about 2,000 ha (Lebaron,1993). By sustaining the provision of design and installation advice and low cost credit, the technology has spread within a localised region of the country where the physical and economic conditions are appropriate. Without this government support it is unlikely that the technology would have spread as it has.

Egan (1997), describing the introduction of treadle pumps by the NGO International Development Enterprises, lays great emphasis on the importance of an effective marketing programme based in and funded by the private sector. An effective marketing network comprises manufacturers, retailers and field technicians capable of demonstrating and installing pumps. Egan identifies the following characteristics of the technology and marketing network as prerequisites for sustainable adoption of the pump technology by smallholders:

- Pumps must be low-cost
- Pumps should be targeted at individual farmers rather than groups
- Not given as a "free gift" to farmers
- The technology should provide a high return to investment. Farmers should be able to recoup investment in less than a year and the equipment should operate for at least 5 times the payback period.
- The pump should be manufactured locally
- Pumps should be manufactured and maintained by the private sector
- Advertising, coupled to a dealer network equipped to meet demand, is essential.

Not all of these characteristics apply to the introduction of different irrigation technologies. They cannot be applied to the introduction of larger communal schemes where farmers draw water from a conveyance network and the approach plays down the role of credit and the use of imported equipment where this may be cheaper than local manufacture. However, the example underscores the importance of co-ordinated actions by the private sector in promoting even a "simple" technology such as the treadle pump.

Zimbabwe appears to be the only low-income developing country pursuing a national policy to provide public irrigation infrastructure promoting use of pressurised irrigation technology by smallholder farmers. The Government has promoted draghose irrigation in the face of severely limited water supply. Typically, in response to requests for assistance from farmers, AGRITEX will produce designs, and layout systems, providing pumps, mains and laterals. Farmers must purchase the field irrigation equipment themselves. The extension services provide good support.

Substantial legislative and financial support from government and the private sector are evident in every country where smallholders have adopted modern systems. They appear to be essential prerequisites for the widespread adoption of modern irrigation technologies by smallholders. Small trials or demonstration plots of the type seen in Sri Lanka (Miller and Tillson, 1989; De Silva, 1995) are unlikely, on their own, to result in widespread adoption of the technology. Support through the provision of competent extension services and some degree of financial support or incentive is also needed.

Marketing and Finance

Improved irrigation technologies involve smallholders in considerable expense. It is commonly believed that farmers will not invest until confident of a rapid return of at least double the investment.

Even the simplest modern irrigation technologies such as the buried clay pipes evaluated in Zimbabwe (Batchelor et al, 1996) and piped distribution networks used widely in China (Yin, 1991) require extra investment by farmers, notwithstanding that subsidies and low-cost loans may be available to cover part of the cost. Table 7 provides a summary of the equipment costs reported in the literature.

To secure a return on investment, equipment must be used to irrigate a high-value marketable crop. Where schemes are reliant on local market outlets there is a significant risk of over-production and consequent drop in prices. In eight case-studies of small-scale irrigation schemes in Africa using very simple technologies - commonly low-lift pumping from surface or shallow groundwater with piped distribution and surface irrigation in small field plots - Carter (1989) reports marketing problems in four cases. Poor technical design and inadequate agricultural extension services were also identified as factors constraining production on half of the schemes studied.

In the gravity sprinkler systems established in Guatemala, local urban centres provide an adequate market and there is no evidence of over-supply and price reductions. However, Keller (1990) reports that attempts to transfer the technology to Ecuador have faced difficulties. There the systems have functioned satisfactorily but farmers use them to irrigate low-value crops, thereby jeopardising the financial viability of the systems.

Cyprus, Israel, Jordan and Egypt all have well-developed internal markets and are also well placed to export to Europe and the Gulf. These markets have contributed to the

36

development of commercial, high-input farming systems, even amongst smallholders. In these farming systems modern irrigation is only one of a range of technologies used to increase the productivity of land, labour and water resources.

The provision of low-cost credit, often coupled with some degree of direct subsidy, is common to all the cases where there was widespread adoption of a modern irrigation technology by smallholders. Credit requires the support of either public or private sector agencies.

Table 7   Reported Costs of Capital Equipment

Equipment Type and Crop	Country	Reference	Cost $US / ha [1]
Hand-move laterals - veg/orchards	Israel	Melamed, 1989	1,400 – 1,600
Drip – orchard	Israel	Melamed, 1989	1,500
	Pakistan	Moshabbir et al, 1993	800
Drip – vegetables	Israel	Melamed, 1989	3,000
	Israel	Regev et al, 1990	4,700[2]
	Jordan	Van Tuijl, 1993	1,000
	India	Suryawanshi, 1995	1,300
	China	Qiu, 1992	4,000
Low technology drip using movable laterals	Nepal	Polak et al, undated	250
Mini sprinklers – vegetables	Israel	Melamed,1989	3,100
Mini sprinklers – orchard	Israel	Melamed, 1989	2,200
Rain-gun sprinkler	Pakistan	ISM/R, 1993	500
Low technology gravity sprinklers	Guatemala	Lebaron 1993	150 – 2,400

1.   All costs are for locally manufactured equipment with the exception of China

2.   Includes cost of main canal lining and farm storage ponds

5 Potential for the African Region

The need for increased food production in Africa has been previously identified. Irrigation has an essential role to play in supplementing rain-fed production.

Appendix 2 includes examples of the introduction of modern technology in Africa. None the less, failures of modern irrigation methods on smallholder developments are not uncommon in Africa. Detailed preliminary investigations are needed to match technologies to local circumstances.

This Chapter reviews the relevance and short-comings of national development indicators as a proxy for identifying potential. As previously indicated, many factors determine whether modern irrigation methods may be adopted. Most of them cannot easily be quantified, for example: the skills and attitudes of farmers; training and extension advice; marketing opportunities; the presence and market orientation of manufacturers and suppliers; the availability of credit and other agricultural inputs. Yet, an overall view of likely potential in a country could be of use to planners.

- **Water shortage.** Shortage of water is an overriding reason for governments actively to promote the use of modern irrigation technologies by farmers. Table 8 shows African countries ranked according to water scarcity per capita, on the basis of data from FAO and World Bank. The figures are derived from an estimate of total national water resources divided by the population, and thus can give no indication of regional variations, which may be extreme. It is pointed out that water scarcity in Table 8 is measured both in terms of a so-called "global" resource (including water flowing in international waterways), and the "internal" resource (water originating within national boundaries). In some cases there will be large differences between the two measures, for example, in the case of Egypt.

- **Agricultural sector performance**. Column 5 in Table 9 shows that food production per capita grew by 1% or more in only 6 African nations, and declined in 24 during the period 1979 – 1992. At present rates of progress, regional deficiencies in food are unlikely to be met from the production of neighbouring countries.

  The poor performance of the agriculture sector in much of the continent suggests that the introduction of innovative new technologies might be questionable. However, it has been shown that in favourable circumstances, simple production packages based on irrigation can be effectively promoted by the private sector, in some cases bypassing the difficulties faced by under-resourced public sector extension services.

- **Industry**. Except for the simplest gravity-fed applications, modern technologies require some basic design, spare parts and technical support for e.g. pumps and distribution equipment. Locally manufactured items, provided they are of a basic quality, have the advantage that they are likely to be supported by local skills whereas there may be problems with imported equipment. Nations with a developing industrial base will have more of the skills and resources needed both to produce and to maintain newer technologies.

Column 6 in Table 9 indicates the extent of industrialisation in African States. Zimbabwe derives some 30% of its GDP from manufacturing, whereas the comparable figure for Rwanda is only 3%. The figures for Morocco, Algeria and Niger are 17%, 11% and 7% respectively. Modern irrigation methods have been adopted by smallholders to a greater or lesser extent in Zimbabwe, Morocco and Algeria. Based on the uptake of technologies by particular nations, it is suggested that the introduction of new methods in developing countries where the manufacturing base supplies less than 10% of GDP, could be problematic.

- **Urbanisation and markets**. High levels of urbanisation indicate the existence of local markets more likely to demand fruit and vegetables, also a possible shortage of labour or higher labour costs. In these circumstances, modern irrigation methods may be financially attractive. Mauritania is an example of a country having a sparse population and a high level (50%) of urbanisation – 83% of the urban population lives in the capital city. Modern irrigation methods might be considered attractive by entrepreneurial farmers on the peri-urban fringe, provided water was available.

Countries use different criteria to define urban settlement. However, given this variability, Column 7 in Table 9 shows that urbanisation ranges from as little as 6% in Rwanda and Burundi (both with very high population densities), to 54% in Tunisia and 57% in Algeria. Countries at the lower end of the range, say below 20%, probably contain few large regional centres with large markets and equipment dealers.

Based on the statistics of Tables 8 and 9, those African States where greater potential for the promotion of appropriate modern irrigation technologies may exist are listed in Table 10. The identification of a national potential is clearly a coarse indicator because there will be wide regional differences characteristic of developing economies. There may also be countries not thus identified, in which selected regions may offer many of the conditions for the uptake of modern technologies. Table 10 therefore merely provides a starting point for further investigations.

Table 8 **African States Ranked by Internal Water Scarcity with Other Nations for Comparison**

Country	Pop. Density person/km^2	Global Water Scarcity m^3/ head	Internal Water Scarcity m^3/ head
AFRICA			
Egypt	55	1,252	33
Libya	3	123	123
Mauritania	2	5,142	180
Niger	7	3,674	396
Tunisia	51	464	417
Algeria	11	544	529
Burundi	222	580	580
Somalia	14	1,487	661
Kenya	47	1,104	739
Rwanda	291	834	834
South Africa	33	1,233	1,105
Morocco	59	1,145	1,145
Sudan	11	5,628	1,279
Zimbabwe	28	1,818	1,282
Malawi	92	1,725	1,614
Burkina Faso	37	1,742	1,742
Ghana	71	3,140	1,788
Uganda	87	3,201	1,891
Benin	46	4,918	1,963
Mauritius	552	1,993	1,993
Botswana	2	10,187	2,010
Nigeria	117	2,581	2,037
Ethiopia	44	2,059	2,059
Chad	5	6,955	2,426
Tanzania	31	3,085	2,773
Gambia	98	7,401	2,775
Togo	70	2,993	2,868
Swaziland	49	5,409	3,125
Senegal	41	4,859	3,256
Namibia	2	30,333	4,133
Cote d'Ivoire	43	5,639	5,566
Mali	8	9,558	5,735
Mozambique	19	13,911	6,440
Zambia	12	12,614	8,721
Angola	9	17,238	17,238
Cameroon	27	20,822	20,822
Zaire	18	23,947	21,973
Madagascar	24	23,561	23,561
Guinea	26	34,764	34,764
Sierra Leone	61	36,347	36,347
CAR	5	43,586	43,586
Liberia	30	78,885	68,004
Congo	7	330,684	88,235
Gabon	5	127,825	127,825
OTHER STATES REVIEWED			
Jordan	44	N/a	405
Israel	243	N/a	431
Cyprus	80	1,253	1,253
India	269	2,094	1,862
China	122	2,420	2,420
Sri Lanka	264	2,586	2,586
Guatemala	89	N/a	11,959

Sources: FAO, 1995b; World Bank, 1994

Table 9   Selected Development Statistics Ranked By GNP/Capita

Country	Main irri. Crop	GNP/head $US 1992	Av ann growth % 1985-92	Food Prod. per capita Growth 1979-92	Manufact. % GDP 1992	% Urbaniz. 1992
1	2	3	4	5	6	7
AFRICA						
Rwanda	Sweet potato	80	-6.6	-2.2	3	6
Mozambique	Rice	90	3.8	-2.1		33
Ethiopia		100		-1.3	3	13
Tanzania	Rice	140	0.8	-1.2	8	24
Burundi	Rice	160	-0.7	0	12	7
Sierra Leone	Rice	160	-0.4	-1.2	2	35
Malawi	Sugarcane	170	-0.7	-5	14	13
Chad		180	0.7	0.3	16	21
Uganda		190	2.3	0.1	7	12
Madagascar	Rice	200	-1.7	-0.16		26
Niger	Rice	230	-2.1	-2	7	22
Kenya	Vegetables	250	0.0	0.1	11	27
Mali	Rice	250	1.0	-0.9	9	26
Nigeria	Rice	280	1.2	2	7	38
Burkina Faso	Rice	300	-0.1	2.8	21	25
Togo	Sugarcane	320	-2.7	-0.7	9	30
Gambia	Rice	330	0.5		7	25
Zambia	Wheat	350	-1.4	-0.8	23	43
Benin	Rice	370	-0.8	1.8	7	41
CAR	Rice	370	-2.7	-1.1		39
Ghana	Rice	410	1.4	0.3	8	36
Mauritania	Sorghum	480	0.2	-1.5	12	52
Zimbabwe	Wheat	500	-0.5	-3.3	30	31
Guinea	Rice	520	1.3	-0.5	5	29
Senegal	Rice	600	-0.7	-0.2	14	42
Cote d'Ivoire	Sugarcane	610	-4.6	0.1	26	43
Congo		620	-2.9	-0.5	7	58
Cameroon	Rice	680	-6.9	-1.7	12	44
Egypt	Berseem	720	1.3	1.4	15	45
Swaziland	Sugarcane	1,100	-1.2			
Morocco	Grain	1,140	1.2	2.3	17	48
Algeria	Vegetables	1,650	-2.5	0.9	11	55
Tunisia	Fruit/grape	1,790	2.1	1.4	20	57
Namibia	Maize	1,970	3.3	-2.5	9	36
Botswana	Vegetables	2,800	6.6	-3.1	4	30
South Africa	Pasture	3,040	-1.3	-2.1	23	50
Mauritius	Sugarcane	3,150	5.8	0.8	22	41
Gabon	Rice	3,880	-3.7	-1.2	11	49
Liberia	Rice	c	N/a	N/a	N/a	N/a
Somalia	Maize	c	-2.3	-6	N/a	N/a
Sudan	Cotton	c	-0.2	-2.2	N/a	N/a
Zaire	Sugarcane	c	-1.0	N/a	N/a	N/a
Libya		d	N/a	N/a	N/a	N/a
Angola	Sugarcane	f	-6.8	N/a	N/a	N/a
OTHER STATES REVIEWED						
India		320	2.9	1.6	18	27
China		530	7.8	2.9	37	29
Sri Lanka		640	2.9	-2.2	16	22
Guatemala		1,200	0.9	-0.8		41
Jordan		1,440	-5.6	-0.5	14	71
Cyprus		10,260	4.6	N/a	N/a	N/a
Israel		14,530	2.3	-1.1	N/a	91

c. Estimated to be low income ($725 or less).   d. Estimated to be upper-middle income ($2896 – $8956).
f. Estimated to be lower-middle income, ($726 - $2895)
Sources: FAO, 1995b; World Bank ,1994

Table 10   **African States with Greater Potential for Use of Modern Irrigation by Smallholders**

Country	Pop. Density person/km²	Internal Water Scarcity m³/head	GNP/head 1992 $US	Food prod. Per capita % Annual growth '79-92	Manufact. % GDP 1992	% Urbaniz. 1992
South Africa	33	1,105	2,670	-2.1	25	50
Egypt	55	33	720	1.4	15	45
Zimbabwe	28	1,282	580	-3.3	30	30
Senegal	41	3,256	600	-0.2	14	42
Mauritania	2	180	530	-1.5	11	50
Ghana	71	1,788	450	0.3	9	35
Kenya	47	739	310	0.1	12	25
Zambia	12	8,721	350	-0.8	23	43
Nigeria	117	2,037	280	2.0	7	38

6 Summary of Findings

This study has identified conditions relating to the availability of water, institutional support and economic opportunity which dispose smallholders to adopt and sustain modern irrigation methods. Physical factors such as climate, soil type and topography determine what irrigation method is appropriate, irrespective of whether the farmer is a smallholder or a large-scale commercial grower, and for that reason they are not referred to directly in this summary.

The following key issues have been identified.

1. **The technology must offer the farmer sufficient financial return or a reduction in labour demand, to justify the capital investment.**

A rule-of-thumb suggests that farmers will be attracted to an innovation if it provides two to three times the returns which would be achieved without the investment. Provided there is an assured market for high-value crops (see point 2) returns on investment in appropriate irrigation equipment will be high when:

- Water is costly:
  Farmers are operating their own pumps and wells, or paying water charges to a supply authority.

- Water is scarce:
  Farmers may either extend the proportion of their holding which is irrigated or get a better output from the same area by more closely meeting crop demands.

- Labour is scarce:
  The more sophisticated technologies offer the greatest potential for saving labour. Less sophisticated equipment, which is more appropriate to the smallholder, such as piped supply for surface irrigation, offers more limited scope for saving labour. Irrigation using draghose sprinklers may be as labour-intensive as surface irrigation methods. Labour savings may therefore be a less attractive incentive in smallholder farming.

2. **Farmers need to grow high-value crops for an assured market in order to cover the costs of the equipment.**

- In almost every case reviewed, modern irrigation equipment is used to irrigate high-value cash crops marketed off the farm. Modern irrigation equipment is seldom used by smallholder farmers to irrigate basic grains or other subsistence crops. Modern methods enable farmers to supplement basic food production and increase earnings from a relatively small area, perhaps during the dry season when food production is not possible. This extra income can provide real benefits to the household.

- Farmers focused on subsistence needs will not take up new technology as readily as those with a more commercial orientation. The desire must be strong to increase production and earnings through irrigation and use of other inputs.

43

- Farmers must have access to markets and supply of agricultural inputs. Fuel cost and supply, road infrastructure and transport costs and proximity to markets will all affect the profitability of irrigated cash crops and will influence farm budgets.

3. **Increasing national or regional water shortage is an important factor motivating governments actively to promote the use of modern irrigation technologies.**

- Volumetric-based charging for water has not been adopted in any of the developing countries considered in this review. In the near future, water charges are unlikely to be a practical means of encouraging farmers to adopt more water-efficient technologies. However, where farmers pay for pumping there is an incentive to operate systems efficiently and to consider alternative methods of irrigation.

- Where governments aim to promote more efficient agricultural water use, incentives must be offered to the farmer. Public infrastructure projects delivering pressurised supply at the farm turnout, price subsidies on equipment and low cost credit are possible means of achieving this.

4. **Government must enact policies promoting the technologies for the smallholder, making it attractive to manufacturers and dealers to develop and promote appropriate irrigation technologies for smallholders.**

- Open up markets:
  - Assist smallholders to gain access to export markets; reduce export tariffs.

- Encouragement of local industry and dealers:
  - Define the balance between equipment imports and local manufacture
  - Set appropriate tariffs on raw materials and finished equipment
  - Encourage local manufacture through tax and other material incentives
  - Price subsidies, if necessary, to promote early uptake

- The private sector has an important role in the supply of equipment.
  Government must facilitate their role through enlightened policies and reduced bureaucracy. The suppliers will need to work closely with the farmers and provide a good after-sales service to instil confidence and ensure sustainability.

- Land tenure rights:
  Define or confirm a legal framework of land tenure, reinforcing traditional rights where appropriate, which will serve as incentive for smallholders to invest in land improvement and irrigation.

- Define credit policies:
  The provision of credit or subsidy, from the public or private sector, has contributed to the successful uptake of technologies in those countries that have now moved away from surface irrigation.

5. **Suitable systems must be relatively cheap and straightforward to operate and maintain.**

- No single type of modern irrigation technology is universally appropriate. Technologies must:
  - Permit cost recovery for the farmer within one to two years
  - Be suitable for use on small and irregular shaped plots
  - Require only simple maintenance skills and equipment
  - Have low risk of component failure
  - Be simple to operate
  - Be durable and reliable - able to withstand rough and frequent handling without serious damage.

Existing technologies that best meet these criteria include:
  - Piped distribution networks including portable, layflat hose
  - Low technology, gravity sprinklers
  - Pressurised bubbler (only suited to orchard crops)
  - Draghose sprinklers

- Systems depending on close agreements between numbers of farmers for efficient operation, or requiring field equipment to be shared by individuals, are unlikely to be sustainable.

- Micro-irrigation technologies are conventionally regarded as too complex to be successful amongst smallholders as they require a good understanding of crop: soil: water interactions and high levels of maintenance. However, India and China have established national manufacturing capacity and are promoting micro-irrigation technologies for smallholders with some success. These programmes lay emphasis on 'simple systems' with no reliance on automatic control or other labour-saving devices.

6. **Farmers require effective technical support in the initial years of adopting an innovation, when they are engaged in a learning process with direct consequences for their income and financial situation. In some cases, the penalty for failure may be ruin and the loss of livelihood.**

Unless farmers are trained in the correct techniques for irrigated cropping systems, the returns they achieve will be sub-optimal.

- Farmers will need to know when, and how much water and other inputs to apply to crops, as well as how to overcome common operational and maintenance problems. They will also need ready access to spare parts for the equipment.

- Staff of government agricultural services may not be specialists in irrigation agronomy. If suitable specialists in public service are not available, experts from the private sector will be needed to advise on cropping; system design; installation; operation and maintenance, possibly working under contract to government agencies.

- Trial and demonstration plots can be effective in promoting a technology amongst smallholders but these must form part of a wider package of support provided by specialist advisers.

7. **Individual, communal and joint state/farmer-owned and operated schemes are all found, and each offer advantages and disadvantages. The preferred system will depend on local criteria. Generalised policies should not be imposed from outside.**

- Where considerable investment has to be made to develop a water source then joint state/farmer operation or the transfer of all infrastructure to a farmer group are possible options, the choice being determined by the management skills of the farmers and the scale of the infrastructure. Farmer-managed schemes can operate successfully but designers must ensure that operational and maintenance requirements are consistent with farmers' interests and abilities. In the current climate of public sector retrenchment, farmer-managed schemes are likely to be the favoured approach of many governments.

- Farmer managed schemes, regardless of the method of irrigation, require competent leadership to be effective. The introduction of improved irrigation methods involves quite radical changes to traditional practices. The initiative for change may come from the group leader or from group members. In either event, the implementation of change requires strong leadership. Groups that are barely managing to co-operate for surface irrigation are unlikely to achieve success with improved methods.

- Schemes where the state retains responsibility for the supply and distribution of water, allowing individual farmers to take water on demand, avoid the problems of establishing and maintaining farmer groups. However, construction costs are higher to provide greater conveyance capacities and on-line storage and farmers must pay for the operation and maintenance of the supply network. Such schemes may not be sustainable in most of the less developed countries because of the high capital and operating costs and poor revenue collection.

- Where a water source can be exploited by an individual farmer the potential problems of joint ownership or farmer co-operation for system O & M can be avoided. Developments by individual farmers may be suited to land immediately adjacent to rivers or lake sides or on land underlain by a shallow aquifer, (no more than 6 or 7 m lift) where relatively low-cost suction lift pumps - manual or motorised - can be used. In the absence of a rural electricity supply, small petrol, diesel or kerosene-driven centrifugal pumps allow irrigation of areas as large as 2 to 3 ha but they are relatively expensive. The financial return from such an area should cover the investment. Treadle pump technology may be appropriate for farmers irrigating smaller plots of up to 0.4 ha. Whatever water lifting technology is used, it must be coupled to appropriate distribution and in-field equipment to make optimum use of the water that is obtained.

46

7 The Promotion of Modern Irrigation Technology

This survey indicates that the "ideal" environment for introducing new methods would have the following characteristics:

A smallholding, with easy access to a major market (perhaps a peri-urban or rural area near a large urban area), and a reliable transport system. Access to credit, and appropriate - that is, low cost, robust - equipment readily available. Good technical advice and support available to the farmer. Security of tenure for the farmer with a local water source (which may be limited in volume). Basic needs are met from rainfed agriculture or irrigation of grain crops, probably during the wet season. The farmer has a desire to increase family income and reduce labour costs and the government is keen to promote new technology and provide the right policy framework to facilitate its uptake.

These ideal characteristics may not all be present in every situation but they summarise the main factors present where new irrigation methods have been adopted. Projects aimed at promoting new methods should be preceded by an investigation to determine whether the existing conditions are suitable or to define what measures are needed to ensure appropriate conditions are established.

It will be necessary to investigate the following issues:

The prevailing financial status, area of holding and form of land tenure of typical smallholders, including:
- level of possible investment in equipment
- current expenditure on labour, water and other inputs
- relative importance of subsistence and commercial cropping and crop types
- willingness to invest in land improvements
- availability of credit

The marketing opportunities, including:
- ease of access to local and export markets and expected prices
- quality requirements for marketing
- opportunities/requirements for co-operative ventures

The present use of 'improved' agricultural practices:
- inputs and/or equipment
- knowledge and availability of other mechanical equipment

Physical characteristics:
- water source
- soil type
- topography
- availability and cost of fuel/energy for pumping

Support services:
- agriculture

47

- dealers
- manufacturers

Where possible, the experience of smallholders elsewhere in the country using the same technology should be reviewed to ensure that the technology is sustainable. The review should identify whether additional support services are needed to overcome specific problems. Where no previous experience is available within the country, effort should be made to find the nearest equivalent example of the intended technology being used by smallholders. Projects should be based on positive interest and proposals from farmers rather than "handed-down" from outside agencies.

Procedures exist for improving the identification of smallholder irrigation projects using traditional methods (ICID 1996, Chancellor and Hide 1997, both with DFID support). The development and field testing of a formalized procedure, perhaps based around a checklist, would assist in the task of identifying sustainable developments based on modern irrigation technologies.

Low cost credit is a prerequisite for many smallholders to invest in all but very simple, low cost equipment. Farmers must be clear about the conditions attached to formal loans. It may be necessary for governments to offer inducements to banks to extend credit either to individual or to groups of smallholders, as repayment rates are traditionally poor. Alternatively, specialised credit provision through national NGOs such as the Smallholder Irrigation Scheme Development Organisation (SISDO) in Kenya, may be a more effective means of providing and controlling loans. It is unlikely that dealers and merchants will be willing to offer credit directly to individual smallholders, given the risk of default, but they may do so to established groups.

Sustainable adoption of even simple technologies such as low technology sprinklers or low-lift pumps and layflat hose, requires that equipment be available from local dealers capable of dealing with the smallholder. Where large suppliers are already dealing with commercial growers they may require training to change their presentation and marketing approach to serve the smallholder sector. Such training may be achieved by collaborative action between donor agencies and the private sector suppliers.

Government policies must provide a supportive environment. This should include support to private sector suppliers or manufacturers in the form of tax incentives or a favourable business climate. Donors could play an important role in promoting the uptake of new methods by advising on policy and providing financial support and technical assistance.

In most cases it is too much to expect smallholders to make the transition from rainfed farming or traditional surface irrigation to sophisticated micro-irrigation systems in a single step. A phased development that first introduces simpler technologies such as low pressure piped distribution and layflat piping or the use of draghose sprinklers is likely to be more successful. At the early stages of an irrigation development farmers must learn a large number of new technical, financial and management skills, of which the operation and maintenance of the irrigation system is only a small part. Robust and 'tolerant' technologies are required. Subsequently, when farmers are familiar with the many other components of a new production system it may be appropriate to introduce more advanced irrigation equipment which offers greater labour and water savings. Designers of distribution systems should hold this possibility in mind when selecting pump capacities, pipe dimensions, pipe layouts, etc.

It is essential to provide smallholders with knowledge of irrigation technologies that are appropriate to their conditions. Equipment demonstrations that remain within the confines of state demonstration farms or research centres are often ineffective, as farmers have little opportunity for hands-on experience and little confidence that the operating conditions truly reflect their own conditions. There is therefore a need for state agencies to work in close collaboration with equipment suppliers to establish demonstration sites, working with, and supporting, progressive farmers. Such sites should be monitored to provide data on operating costs, labour and water inputs, crop yields, etc for comparison with the surrounding farming systems. However, the primary function is demonstration, allowing farmers to see systems operating and to evaluate them by their own criteria.

Farmers require security of tenure before investing in relatively expensive irrigation equipment. Many smallholders, particularly in Africa, do not have formal tenure, even though in practice they may have adequate security. The situation must be clearly understood and agreed by all concerned parties before embarking on the promotion of modern methods. Where government-funded projects are established, the rights and responsibilities of farmers on the project must be well defined. Donors can provide support in the establishment of the right legislative or land registration structure, in building awareness, improving technical capability and providing demonstration plots.

Farmers must achieve good early returns to investment. Such returns can be achieved only by a package of measures with which traditional farmers may not be familiar and there is therefore a need for advice from specialised irrigation agronomists. Any externally supported project must embrace the full range of measures discussed in this report.

8 References and Bibliography

This section is arranged as follows: the first part provides an alphabetical listing of all references cited in the main body of this report. The second part lists general references relating to irrigation technologies and is followed by material that is specific to a region or individual country.

References

Abbott, J. S. 1988.
Micro-irrigation World Wide Usage. Report by Micro Irrigation Working Group.
ICID Bulletin, Jan 1988. Vol 37 (1) 1-12.

Abbott, J. S. 1984.
Micro-irrigation - World Wide Usage. ICID Bulletin, Jan 1984. Vol 33 (1) 4-9.

Batchelor, C., Lovell, C. and Murata, M. 1996.
Water Use Efficiency of Simple Subsurface Irrigation Systems. In: Proceedings of 7th International Conference on Water and Irrigation. 13 -16 May 1996. Tel Aviv, Israel. pp 88-96.

Battikhi, A.M., and Abu-Hammad, A. H. 1994.
Comparison between the efficiencies of surface and pressurised irrigation systems in Jordan. Irrigation and Drainage Systems, Vol 8 109-121.

Bedini, F. 1995.
Making Water: An Appraisal of 'Jua Kali' Sprinklers in Kenya. Terra Nuara, Nairobi.

Bucks, D. A. 1993.
Micro-irrigation - Worldwide usage report. In: Proceedings of Workshop on Micro-irrigation, Sept 2 1993. 15th Congress on Irrigation and Drainage, The Hague. ICID. pp11-30.

Carter, R. 1989.
NGO Casebook on Small Scale Irrigation in Africa. FAO, Rome.

Chancellor F.M. and Hide, J.M. 1997.
Smallholder Irrigation: Ways Forward. Report OD 136, HR Wallingford, Wallingford, UK.

Chen Dadiao, 1988.
Sprinkler Irrigation and Mini-irrigation in China. In: Proceedings of. International Conference on Irrigation System Evaluation and Water Management.   Wuhan Univ. Vol 1.pp 288-296.

Dalvi, V. B., Satpute, G. U., Pawade, M.N. and Tiwari K. N. 1995.
Growers' experiences and on-farm micro-irrigation efficiencies. In: Proceedings of 5th International Micro-irrigation Congress, April 2-6, 1995, Florida. ASAE. pp 775-780.

De Silva, C. S. 1995.
Drip irrigation with agrowells for vegetable production in Sri Lanka. In: Micro-irrigation for a Changing World. Proceedings of the 5th International Micro-irrigation Congress, April 2-6 1995. F.J. R. Lamm (ed). ASAE. pp 949-954.

Egan, L. 1997.
The Experiences of IDE in mass marketing of small scale affordable irrigation devices. Paper presented at FAO/IPTRID Sub-regional Workshop on Irrigation Technology Transfer in Support of Food Security. Harare, 14-17 April, 1997.

FAO, 1996.
Proceedings of World Food Summit, Rome, 13-17, November 1996.

FAO, 1995 a.
Water development for food security. Draft document prepared for World Food Summit. WFS/96/TECH/2.

FAO, 1995 b.
Irrigation in Africa in Figures. Water Reports No. 7. FAO, Rome. ISSN 1020-1203.

FAO/IPTRID, 1997.
Workshop on Irrigation Technology Transfer in Support of Food Security. Harare, 14 - 17 April 1997.

Field, W. P. 1990.
World irrigation. Irrigation and Drainage Systems, Vol. 4 91-107.

Hanbali, U., Tleel, N. and Field, W. P. 1987.
Mujib and Southern Ghors Irrigation Project. In: Proceedings of 13th Congress of International Commission on Irrigation and Drainage, Sept. 1978. 183-195.

Hargreaves, G. H. 1996.
Making Irrigation more Profitable and Competitive in the Developing Countries. ICID Journal Vol. 45 (2) 13-20.

Hillel, D. 1989.
Adaptation of modern irrigation methods to research priorities of developing countries. In: Le Moigne, Barghouti and Plusquellec (eds). Technological and Institutional Innovation. World Bank Technical Paper No. 94. 88-93.

Hinton, R. D., El Ouosy, D.' Talaat, A. A., and Khedr, M. 1996.
The performance of the El Hammami irrigation pipeline, Egypt. Implication for design and management. Report OD 135. HR Wallingford, UK.

Hlavek, R. 1995.
Selection Criteria for Irrigation Systems. ICID, New Delhi.

Hoffman, G.J. and Martin, D.L. 1993.
Engineering systems to enhance irrigation performance. Irrigation Science, Vol 14, (2) 53-63.

Holsambre, D. G. 1995.
Status of drip irrigation systems in Maharashtra. In: Micro-irrigation for a Changing World. Proceedings of the 5th International Micro-irrigation Congress, April 2-6 1995. F.J. R. Lamm (ed). ASAE. pp 497-501.

Hull, P. J. 1981.
A low pressure irrigation system for orchard tree and plantation crops. The Agricultural Engineer 362: 55-58.

International Commission on Irrigation and Drainage, 1996.
Checklist to Assist Preparation of Small-scale Irrigation Projects in sub-Saharan Africa.

Irrigation Systems Management Research Project, 1993.
Improving on-farm water use and application. Final Report. Water Resources Research Institute, National Agricultural Research Centre and Pakistan Agricultural Research Council

Keen, M. 1991.
Drip-trickle irrigation boosts Bedouin farmers' yields. Ceres No. 130. Vol 23 (4) July-August 1991.

Keller, J. 1990.
Modern irrigation in developing countries. Proceedings of 14th International Congress on Irrigation and Drainage. Rio de Janeiro, ICID. April - May 1990.

Keller, J. 1988.
Taking advantage of modern irrigation in developing countries. In: Drought, water management and food production: Conference proceedings, Agadir, Morocco, November 21-24, 1985. Mohammedia, Morocco: Fedala. pp.247-260.

Keller, J. and Bliesner, R. D. 1990.
Sprinkler and Trickle Irrigation. Van Nostrand Reinhold, New York.

Kezong, X. 1993.
Effects of water saving irrigation techniques in some areas of China. In: Proceedings of 15th Congress on Irrigation and Drainage, The Hague. Vol 1-G 63-73. ICID.

Lebaron, A. D. 1993.
Profitable small-scale sprinkle irrigation in Guatemala. Irrigation and Drainage Systems, Vol 8 13-23.

Lebaron, A., Tenney, T., Smith B. D., Embry, B. L. and Tenney, S. 1987.
Experience with Small-Scale Sprinkler System Development in Guatemala: An
Evaluation of Program Benefits. Water Management Synthesis II Report 68. USAID.

Le Moigne, G. 1989.
Overview of technology and research issues in irrigation. In: Le Moigne, Barghouti
and Plusquellec (eds). Technological and Institutional Innovation. World Bank
Technical Paper No. 94. World Bank, Washington.

Le Moigne, G., Barghouti, S. and Plusquellec H. 1989.
Technological and Institutional Innovation. World Bank Technical Paper No. 94.
World Bank, Washington.

Louw, A. 1996.
Agricultural Research Council, Institute for Agricultural Engineering, Silverton, S.
Africa. Personal communication.

Lyle, W. M. and Bordovsky, J. P. 1983.
LEPA irrigation system evaluation. Trans ASAE 26:776-781.

Manig, W. 1995.
Suitability of Mechanised Irrigation Systems for Peasant Farmers in Developing
Countries. ICID Journal, 44 (1) 1-10.

Melamed, D. 1989.
Technological Developments: The Israeli Experience. In: Technological and
Institutional Innovation in Irrigation. Le Moigne G., Barghouti S. and Plusquellec H.
(eds). World Bank, Technical Paper No. 94. World Bank, Washington.

Miller, E. and Tillson, T. J. 1989.
Small Scale Irrigation in Sri Lanka: Field Trials of a Low Head Drip System. In:
Irrigation Theory and Practice, Proceedings of the International Conference, Uni.
Southampton, 12-15 September 1989. pp 616-629.

Moshabbir P. M., Ahmad S., Yasin M. and Ahmad M. M. 1993.
Indigenization of trickle irrigation technology. In: Government of Pakistan - USAID
Irrigation Systems Management Research Project; IIMI, Proceedings: Irrigation
Systems Management Research Symposium, Lahore, 11-13 April 1993. Vol.VII. -
Improving on-farm water use and application. pp.79-89.

Nir, D. 1995.
Introduction of pressure irrigation in developing countries. In: Micro-irrigation for
a Changing World. Proceedings of the 5th International Micro-irrigation Congress,
April 2-6 1995. F. J.R. Lamm (ed). ASAE. pp 442-445.

Or, U. 1993.
Why micro-irrigation is not being implemented as it should and what should be
done. In: Proceedings of Workshop on Micro-irrigation, Sept 2, 1993. 15th Congress on
Irrigation and Drainage, The Hague. ICID. pp 91-105.

Polak P., Nanes R. and Adhikari D. no date.
A low cost drip irrigation system: affordable access to water-saving irrigation for small farmers in developing countries. International Development Enterprises, Lakewood, Colorado, USA.

Qiu, W. 1992.
Development of drip irrigation technology in arid areas of China. In: Shalhevet, J. Liu, C. Xu, Y. (Eds.) Water use efficiency in agriculture: Proceedings of the Binational China-Israel Workshop, Beijing, China, 22-26 April 1991. Rehovot, Israel: Priel Publishers pp 252-257.

Rao, D. S. K. 1992.
Community sprinkler system in Sullikere village, Bangalore urban district, South India. In: Abhayaratna, M. D. C., Vermillion, D., Johnson, S., Perry, C. (Eds). Farmer management of groundwater irrigation in Asia: Selected papers from a South Asian Regional Workshop on Groundwater Farmer-Managed Irrigation Systems and Sustainable Groundwater Management, held in Dhaka, Bangladesh from 18 to 21 May 1992. Colombo, Sri Lanka: IIMI. pp 139-151.

Regev, A., Jaber A., Spector R. and Yaron D. 1990.
Economic Evaluation of the Transition from a Traditional to a Modernised Irrigation Project. Agricultural Water Management, 18 347-363.

Reynolds, C., Yitayew, M. and Petersen M. 1995.
Low-head bubbler irrigation systems. Part 1: Design. Agricultural Water Management 29 (1) 1-24.

Rolland, L. 1982.
Mechanized Sprinkler Irrigation. FAO, Irrigation and Drainage Paper No. 35. FAO, Rome.

Rukuni, M. 1997.
Creating an enabling environment for the uptake of low-cost irrigation equipment by small-scale farmers. Paper presented at FAO/IPTRID Sub-regional Workshop on Irrigation Technology Transfer in Support of Food Security. Harare, 14 - 17 April, 1997.

Saksena, R. S. 1993.
Status of micro-irrigation in India. In: Proceedings of Workshop on Micro-irrigation, Sept 2 1993. 15th Congress on Irrigation and Drainage, The Hague. ICID. pp 41-52.

Saksena, R. S. 1995.
Micro-irrigation in India - Achievement and perspective. In: Micro-irrigation for a Changing World. Proceedings of the 5th International Micro-irrigation Congress, April 2-6 1995. F.J. R. Lamm (ed). ASAE. pp 353-358.

Samani, Z., Rojas, H. and Gallardo, G. 1991.
Adapted drip irrigation technology for developing countries. In: Ritter W. F. (ed). Irrigation and Drainage. Proceedings of 1991 national conference. Irrigation and drainage div. ASCE.

Sharma, B.R. and Abrol, I.P. 1993.
Future of Drip and Sprinkler Irrigation Systems in India. In: Proceedings of Workshop on Sprinkler and Drip Irrigation Systems. 8-10 December 1993. Jalgoan, Central Board of Irrigation and Power, New Delhi. pp 21-25.

Shelke, P. P., Singh, K. K., and Chauhan, H. S. 1993.
Socio-economic aspects of use of sprinklers in Sikar District, Rajasthan. In: Proceedings of Workshop on Sprinkler and Drip Irrigation Systems. 8 - 10 December 1993 Jalgaon, Central Board of Irrigation and Power, New Delhi. pp 81-83.

Singh, J., Singh A.K. and Garg, R. 1993.
Present status of drip irrigation in India. In: Proceedings of Workshop on Sprinkler and Drip Irrigation Systems. 8 - 10 December 1993 Jalgaon, Central Board of Irrigation and Power, New Delhi. pp 11-15.

Sivanappan, R. K. 1988.
Cost Benefit ratios and case studies: Unpublished papers on Drip Irrigation in South India.

Suryawanshi, S.K. 1995.
Success of Drip in India: An example to the world. In: Micro-irrigation for a Changing World. Proceedings of the 5th International Micro-irrigation Congress, April 2-6 1995. F.J. R. Lamm (ed). ASAE. pp 347-352.

United Nations Economic & Social Commission for Asia & the Pacific, 1995.
Guidebook to water resources use and management in Asia and the Pacific. Vol one: Water Resources and Water Use. Water Resources Series No. 74. United Nations, New York.

Van Bentum, R. and Smout, I. K. 1994.
Buried pipelines for surface irrigation. Intermediate Technology Publications, in association with WEDC, Loughborough.

Van Tuijl, W. 1993.
Improving Water Use in Agriculture, Experiences in the Middle East and North Africa. World Bank Technical Paper No 201. World Bank, Washington DC.

Winpenny , J.T. 1997.
Managing water scarcity for water security. Discussion paper prepared for FAO.

World Bank, 1994.
World Development Report, Infrastructure for Development. Oxford University Press.

Yin Jie 1991.
The operation management and economic effect of irrigation through plastic flexible hose. ICID Special Technical Session, Beijing, China. April 1991. Vol 1- C. pp 96-101.

Zadrazil, H. 1990.
Dragline irrigation: Practical experience with sugar cane. Agricultural Water Management Vol 17 25-35.

General Bibliography- Economics, Surveys, Equipment Specifications

Abbott, J. S. 1988.
Micro-irrigation - World Wide Usage. Report by Micro-irrigation Working Group. ICID Bulletin, Jan 1988. Vol 37 (1) pp 1-12.

Abbott, J. S. 1984.
Micro-irrigation - World Wide Usage. ICID Bulletin, Jan 1984. Vol 33 (1) pp 4-9.

Batchelor, C. Lovell, C. and Murata, M. 1993.
Micro-irrigation techniques for improving irrigation efficiency on vegetable gardens in developing countries. In: Proceedings of Workshop on Micro-irrigation, Sept 2 1993. 15th Congress on Irrigation and Drainage, The Hague. ICID. pp 31-39.

Bucks, D. A. 1995.
Historical Developments in Micro-irrigation. In: Micro-irrigation for a Changing World. Proceedings of the 5th International Micro-irrigation Congress, April 2-6 1995. F.J. R. Lamm (ed). ASAE. pp 1-5.

Bucks, D. A. 1993.
Micro Irrigation Worldwide Usage Report. In: Proceedings of Workshop on Micro-irrigation, Sept 2 1993. 15th Congress on Irrigation and Drainage, The Hague. ICID. pp 11-30.

Caswell, M. F. 1989.
The adoption of low-volume irrigation technologies as a water conservation tool. Water International, Vol 14, 19-26.

Chancellor F. M. and Hide J. M. 1997
Smallholder Irrigation: Ways Forward. Report OD 136, HR Wallingford, Wallingford, UK.

Hillel, D. 1989.
Adaptation of Modern Irrigation Methods to Research Priorities of Developing Countries. In: Technological and Institutional Inovation in Irrigation. Le Moigne G., Barghouti S. and Plusquellec H. (Eds). World Bank, Technical Paper No. 94. World Bank, Washington.

Hlavek, R. 1995
Selection Criteria for Irrigation Systems. ICID, New Delhi.

Hoffman, G. J. and Martin, D.L. 1993
Engineering systems to enhance irrigation performance. Irrigation Science 14(2) 53-63.

Hull, P.J. 1981.
A low pressure irrigation system for orchard tree and plantation crops. The
Agricultural Engineer 36(2) 55-58.

International Commission on Irrigation and Drainage, 1996
Checklist to Assist Preparation of Small-Scale Irrigation Projects in Sub-Saharan
Africa.

Jurriens, M. 1982.
Surface, Sprinkler and Drip Irrigation. A Review of Some Selection Parameters.
ILRI Internal Communication.

Kay, M. 1983.
Sprinkler Irrigation Equipment and Practice. English Language Book Society,
London. pp 120.

Keller, J. 1990.
Modern irrigation in developing countries. Proceedings of 14th International
Congress on Irrigation and Drainage. Rio de Janeiro, ICID. April - May 1990.

Keller, J. 1988
Taking advantage of modern irrigation in developing Countries. In: Drought, water
management and food production: Conference proceedings, Agadir (Morocco),
November 21-24, 1985. Mohammedia, Morocco: Fedala. pp.247-260.

Keller, J. and Bliesner, R.D. 1990.
Sprinkler and Trickle Irrigation. Van Nostrand Reinhold, New York. pp 652

Lyle, W.M. and Bordovsky, J.P. 1983.
LEPA irrigation system evaluation. Transactions of the ASAE 26(3) 776-781.

Manig, W. 1995.
Suitability of Mechanised Irrigation Systems for Peasant Farmers in Developing
Countries. ICID Journal 44(1) 1-10.

Nardulli, S. 1995.
Treading Water. Ceres No. 156. November - December 1995 Vol 27 (6) 39-43.

Nir, D. 1995.
Introduction of pressure irrigation in developing countries. In: Micro-irrigation for
a Changing World. Proceedings of the 5th International Micro-irrigation Congress,
April 2-6 1995. F.J. R. Lamm (ed). ASAE. pp 442-445.

Polak, P., Nanes, R. and Adhikari,D. (no date).
A low cost drip irrigation system: affordable access to water-saving irrigation for
small farmers in developing countries. International Development Enterprises,
Lakewood, Colorado, USA.

Reynolds, C. Yitayew, M. and Petersen, M. 1995.
Low-head bubbler irrigation systems. Part 1: Design. Agricultural water management 29(1) 1 – 24.

Rolland, L. 1982
Mechanized Sprinkler Irrigation. FAO Irrigation and Drainage Paper No. 35. FAO, Rome.

Samani, Z. Rojas, H. and Gallardo, G. 1991.
Adapted drip irrigation technology for developing countries. In: Ritter W.F. (ed). Irrigation and Drainage. Proceedings of 1991 national conference. Irrigation and drainage div. ASCE.

Shrestha, R. B. and Gopalakrishnan, C. 1993.
Adoption and diffusion of drip irrigation technology: an econometric analysis. Economic Development and Cultural Change. 41(2) 407-418.

Van Bentum, R. and Smout, I. K. 1994.
Burried Pipelines for Surface Irrigation. Intermediate Technology Publications in association with WEDC.

Wolff, P. and Huebener, R. 1994.
Technological Innovations in Irrigated Agriculture. In: Heim F., and Abernethy C. L. (eds). Irrigated agriculture in Southeast Asia beyond 2000: Proceedings of a workshop at Langkawi, Malaysia. 5 - 9 October, 1992. IIMI, Colombo, Sri Lanka. pp 115 - 125.

Zazueta, F. S. 1995.
International Developments in Micro-irrigation. In: Micro-irrigation for a Changing World. Proceedings of the 5th International Micro-irrigation Congress, April 2-6 1995. F.J. R. Lamm (ed). ASAE. pp 214 – 224.

Zilberman D. 1987.
Focus: The Economics of Irrigation Technology choices. In: Irrinews, No 35 Newsletter of the International Irrigation Information Center, Volcani Centre, Israel.

Africa – General

Agodzo, S. K. and Kyei Baffour, 1992.
Technology changes in irrigation and food security in Africa. In: Proceedings of Conference on Advances in Planning, Design and Management of Irrigation Systems as Related to Sustainable Land Use. Leuven, Belgium, September 14-17, 1992. Vol 1. pp 125-135.

China

Backhurst, A. 1995.
Jiangxi sandy wasteland development project. EC/ALA/CHN/9214. Unpublished project profile prepared by Technical Assistance Consultant, Agrisystems, Aylesbury, UK.

Chen Dadiao 1988.
Sprinkler Irrigation and mini-irrigation in China. In: Proc. International Conference on Irrigation System Evaluation and Water Management. Wuhan Univ. Vol 1. pp 288-296.

Kezong, X. 1993.
Effects of water saving irrigation techniques in some areas of China. In: Proceedings of 15th Congress on Irrigation and Drainage, The Hague. Vol 1-G 63-73. ICID.

Qiu, W. 1991.
Drip irrigation technology - an orientation for development of irrigation technology in arid areas of China. In: Proceedings of Special Technical Session, ICID Beijing, China. April 1991. Vol 1-A pp 300-305.

Cyprus

Van Tuijl, W. 1993.
Improving Water Use in Agriculture, Experiences in the Middle East and North Africa. World Bank Technical Paper No 201. World Bank, Washington DC.

Van Tuijl, W. 1989.
Irrigation Developments and Issues in EMENA Countries. In: Le Moigne, G; Barghouti, S and Plusquellec H (1989). Technological and Institutional Innovation. World Bank Technical Paper No. 94. pp 13-22.

Guatemala

Lebaron, A., Tenney, T., Smith, B.D., Embry, B.L. and Tenney, S. 1987.
Experience with Small-Scale Sprinkler System Development in Guatemala: An Evaluation of Program Benefits. Water Management Synthesis II Report 68. USAID.

Lebaron, A. 1993.
Profitable small-scale sprinkler irrigation in Guatemala Irrigation and Drainage Systems 8 (1) 13-23.

India

Chatterjee, P. K. 1993.
Availability of credit for drip irrigation systems in India. In: Proceedings of Workshop on Sprinkler and Drip Irrigation Systems. 8 - 10 December 1993, Jalgaon, Central Board of Irrigation and Power, New Delhi. pp 109-111.

Chauhan, H. S. 1995.
Issues of Standardisation and Scope of Drip Irrigation in India. In: Micro-irrigation for a Changing World. Proceedings of the 5th International Micro-irrigation Congress, April 2-6 1995. F.J. R. Lamm (ed). ASAE. pp 446 - 451.

Dalvi, V.B., Satpute, G.U., Pawade, M.N. and Tiwari, K. N. 1995.
Growers experiences and on-farm micro-irrigation efficiencies. In: Proceedings of 5th International Micro-irrigation Congress, April 2-6, 1995, Florida. ASAE. pp 775-780.

Dua, S. K. 1995.
The Future of Micro-irrigation. In: Micro-irrigation for a Changing World. Proceedings of the 5th International Micro-irrigation Congress, April 2-6 1995. F.J. R. Lamm (ed). ASAE. pp 341-346.

Holsambre, D. G. 1995.
Status of drip irrigation systems in Maharashtra. In: Micro-irrigation for a Changing World. Proceedings of the 5th International Micro-irrigation Congress, April 2-6 1995. F.J. R. Lamm (ed). ASAE. pp 497-501.

Malavia, D.D Khanpara, V.D. Shobhana, H.K. and Golakiya B.A. 1995.
A comparison of irrigation methods in arid and semi-arid western Gujarat, India. In: Micro-irrigation for a Changing World. Proceedings of the 5th International Micro-irrigation Congress, April 2-6 1995. F.J. R. Lamm (ed). ASAE. pp 464-469.

Patil, V.K. and Chougule, A. A. 1993.
Drip irrigation - Indian scenario. In: Proceedings of 15th Congress on Irrigation and Drainage, The Hague. Vol 1-A 15-32.

Rao, D.S.K. 1992.
Community sprinkler system in Sullikere village, Bangalore urban district, South India. In: Abhayaratna, M. D. C.; Vermillion, D.; Johnson, S.; Perry, C. (Eds.), Farmer management of groundwater irrigation in Asia: Selected papers from a South Asian Regional Workshop on Groundwater Farmer-Managed Irrigation Systems and Sustainable Groundwater Management, held in Dhaka, Bangladesh from 18 to 21 May 1992. Colombo, Sri Lanka: IIMI. pp.139-151.

Saksena, R. S. 1995.
Micro-irrigation in India - Achievement and perspective. In: Micro-irrigation for a Changing World. Proceedings of the 5th International Micro-irrigation Congress, April 2-6 1995. F.J. R. Lamm (ed). ASAE.

Saksena, R.S. 1993a.
Sprinkler and Drip irrigation in India - present bottlenecks and suggested measures for speedier development. In: Proceedings of Workshop on Sprinkler and Drip Irrigation Systems. 8 - 10 December 1993 Jalgaon, Central Board of Irrigation and Power, New Delhi. pp 26-37.

Saksena, R. S. 1993b.
Status of micro-irrigation in India. In: Proceedings of Workshop on Micro-irrigation, Sept 2 1993. 15th Congress on Irrigation and Drainage, The Hague. ICID. pp 41-52.

Saksena, R.S. 1992.
Drip irrigation in India: Status and issues. Land Bank Journal, Bombay, India. March 1992.

Sawleshwarker, N.R. 1995.
Application of Micro-irrigation Technology to Major Irrigation Projects. In: Micro-irrigation for a Changing World. Proceedings of the 5th International Micro-irrigation Congress, April 2-6 1995. F.J. R. Lamm (ed). ASAE. pp 550-551.

Sharma, B. R. and Abrol, I.P. 1993.
Future of Drip and Sprinkler Irrigation Systems in India. In: Proceedings of Workshop on Sprinkler and Drip Irrigation Systems. 8 - 10 December 1993 Jalgaon, Central Board of Irrigation and Power, New Delhi. pp 21-25.

Shelke, P.P., Singh, K.K. and Chauhan, H.S. 1993.
Socio-economic aspects of use of sprinklers in Sikar District, Rajasthan. In: Proceedings of Workshop on Sprinkler and Drip Irrigation Systems. 8 - 10 December 1993 Jalgaon, Central Board of Irrigation and Power, New Delhi. pp 81-111.

Singh, J., Singh, A.K. and Garg, R. 1995.
Scope and potential of drip and sprinkler irrigation systems in Rajasthan, India. In: Micro-irrigation for a Changing World. Proceedings of the 5th International Micro-irrigation Congress, April 2-6 1995. F.J. R. Lamm (ed). ASAE. pp 457-463.

Singh, J., Singh, A.K. and Garg, R. 1993.
Present status of drip irrigation in India. In: Proceedings of Workshop on Sprinkler and Drip Irrigation Systems. 8 - 10 December 1993 Jalgaon, Central Board of Irrigation and Power, New Delhi. pp 11-15.

Sivanappan, R. K. 1994.
Prospects of Micro-Irrigation in India. Irrigation and Drainage Systems Vol 8. pp 49-58.

Israel

Keen, M. 1991.
Drip-trickle irrigation boosts Bedouin farmers' yields. Ceres No. 130. Vol 23 (4) July-August, 1991.

Melamed, D. 1989.
Technological Developments: The Israeli Experience. In: Technological and Institutional Innovation in Irrigation. Le Moigne G., Barghouti S. and Plusquellec H. (Eds). World Bank, Technical Paper No. 94. World Bank, Washington.

Or, U. 1985.
Jordan Valley Drip Irrigation Scheme - A model for developing countries. In: Whitehead, E. Hutchinson C. Timmesman B. Varady R. (Eds.), Arid lands: Today and tomorrow. Fort Collins, CO, USA: Westview Press. pp.189-193.

Regev, A., Jaber, A., Spector, R. and Yaron, D. 1990.
Economic Evaluation of the Transition from a Traditional to a Moderized Irrigation Project. Agricultural Water Management, 18 347-363.

Van Tuijl, W. 1993.
Improving Water Use in Agriculture, Experiences in the Middle East and North Africa. World Bank Technical Paper No 201. World Bank, Washington DC.

Yaron, D. and Regev, A. 1989.
Is Modernization of traditional irrigation systems in arid zones economically justified? In: Irrigation Theory and Practice, Proceedings of the International Conference, Uni. Southampton, 12-15 September 1989. pp 201-210.

Jordan

Battikha, A. M. and Abu-Mohammad, A. H. 1994.
Comparison between efficiencies of surface and pressurised irrigation systems in Jordan. Irrigation and Drainage Systems. Vol 8. 109 – 121.

Hanbali, U. Tleel, N. and Field, W. P. 1987
Mujib and Southern Ghors Irrigation Project. Transaction of 13th International Congress on Irrigation and Drainage, Rabat. ICID. Vol 1-A 183 -195.

Or, U. 1993.
Why micro-irrigation is not being implemented as it should and what should be done. In: Proceedings of Workshop on Micro-irrigation, Sept 2 1993. 15th Congress on Irrigation and Drainage, The Hague. ICID. pp 91-105.

Van Tuijl, W. 1993.
Improving Water Use in Agriculture, Experiences in the Middle East and North Africa. World Bank Technical Paper No 201. World Bank, Washington DC.

Pakistan

Ahmad, S., Moshabbir, P. M., Bhatti, A. A. and Yasin, M. 1993.
Design and local manufacturing of raingun sprinkler irrigation systems. In: Government of Pakistan - USAID Irrigation Systems Management Research Project; IIMI, Proceedings: Irrigation Systems Management Research Symposium, Lahore, 11-13 April 1993. Vol.VII. - Improving on-farm water use and application. pp 55-78.

ISMR/R 1993
Irrigation Systems Management Research Project. Final Report, Improving on-farm water use and application. Booklet VIII. Pakistan Agricultural Research Council, Islamabad.

Keller, J. and Burt, C.M. 1975.
Recommendations for Trickle and Sprinkle Irrigation in Pakistan. Unpublished report on a field trip 7 -19 April 1975.

Moshabbir, P. M., Ahmad, S., Yasin, M. and Ahmad, M. M. 1993.
Indigenization of trickle irrigation technology. In: Government of Pakistan - USAID Irrigation Systems Management Research Project; IIMI, Proceedings: Irrigation Systems Management Research Symposium, Lahore, 11-13 April 1993. Vol.VII. - Improving on-farm water use and application. pp.79-89.

South Africa

De Lange, M. 1994.
Small scale irrigation in South Africa. WRC Report No. 578/1/94. Pretoria, South Africa

Sri Lanka

Batchelor, C., Lovell, C. and Murata, M. 1993.
Micro-irrigation techniques for improving irrigation efficiency on vegetable gardens in developing countries. In: Proceedings of Workshop on Micro-irrigation, Sept 2 1993. 15th Congress on Irrigation and Drainage, The Hague. ICID. pp 31-39.

De Silva, C. S. 1995.
Drip irrigation with agrowells for vegetable production in Sri Lanka. In: Micro-irrigation for a Changing World. Proceedings of the 5th International Micro-irrigation Congress, April 2-6 1995. F.J. R. Lamm (ed). ASAE. pp 949-954.

Foster, W.M., Batchelor, C.H., Bell, J.P., Hodnett, M.G., and Sikurajapthy, M. 1989.
Small Scale Irrigation in Sri Lanka Soil Moisture Status and Crop Response. In: Irrigation Theory and Practice, Proceedings of the International Conference, Uni. Southampton,12-15 September 1989. pp 602-615.

Miller, E. and Tillson, T.J. 1989.
Small Scale Irrigation in Sri Lanka: Field Trials of a Low Head Drip System. In: Irrigation Theory and Practice, Proceedings of the International Conference, Uni. Southampton, 12-15 September 1989. pp 616-629.

Zimbabwe

Batchelor, C. 1984.
Drip Irrigation for Small Holders. In: Proceedings of African Regional Symposium on Small Holder Irrigation. University of Zimbabwe, Harare 5-7-Sept 1984. HR Wallingford and University of Zimbabwe. pp115-122.

Batchelor, C., Lovell, C. and Murata, M. 1993.
Micro-irrigation techniques for improving irrigation efficiency on vegetable gardens in developing countries. In: Proceedings of Workshop on Micro-irrigation, Sept 2 1993. 15th Congress on Irrigation and Drainage, The Hague. ICID. pp 31-39.

Lovell, C. J. *et al.* 1996.
Small-scale irrigation using collector wells pilot project - Zimbabwe. Report ODA 95/14, Institute of Hydrology, Wallingford, UK.

Murata, M., Batchelor, C., Lovell, C.J. Brown, M.W., Semple, A.J., Mazhangara, E., Haria, A., McGrath, S.P. and Williams, R.J. 1995.
Development of small-scale irrigation using limited groundwater resources. Fourth Interim Report. Institute of Hydrology, Wallingford, UK. Report ODA 95/5.

Soloman, K.H. and Zoldoske, D.F. 1994.
Establishing irrigation equipment testing in Zimbabwe. In: Cartwright A. (ed) World Agriculture 1994. Sterling Publications, London, UK. pp 96-98.

Stoutjesdijk, J. A. 1989.
Aspects of small-scale irrigation in the southern African region. In: Irrigation Theory and Practice, Proceedings of the International Conference, Uni. Southampton, 12-15 September 1989. pp 182-191.

Watermeyer, 1986.
Are Sprinkler Systems suitable for communal irrigation settlements? Presented at joint Kenya/Zimbabwe workshop on irrigation policy. April 1986.

9 Acknowledgements

The support of DFID for the one-year study is gratefully acknowledged.

Mr Charles Batchelor of the Institute of Hydrology is thanked for his advice and assistance.

Professor Dov Nir of TECHNION – the Israel Institute of Technology, contributed significantly to earlier work carried out within the ODU at HR Wallingford on modern irrigation technologies.

Appendix 1

Uptake of Micro-Irrigation

Appendix 1   Uptake of Micro-Irrigation Technology

The ICID Working Group on Micro-irrigation has carried out three surveys reviewing the usage of micro-irrigation technologies, (Abbott, 1984; Abbott, 1988 and Bucks, 1993). The data obtained relate only to the use of micro- or localised irrigation and exclude information on different types of overhead sprinkler irrigation. The first survey gathered information from selected member nations of the ICID, whilst the last two have sought to collate data from all member countries. The findings of the surveys, shown in Tables A1.1and A1.2, give some indication of trends regarding the extent and rate of expansion of area under different forms of micro-irrigation. However, the values are only approximate guides, based on the information available to members of each ICID National committee. It is particularly notable that 13 of the countries returning data in 1991 reported exactly the same area as in 1986, suggesting that the 1991 survey was simply repeating data reported in 1986.

Japan, Thailand, Austria, Mexico and Italy have shown the greatest increases in area under micro-irrigation but in none of these countries does micro-irrigation account for more than 5% of the total irrigated area.

Cyprus, Israel and Jordan, all in the eastern Mediterranean, stand out in Table A1.2 having between 20% and 70% of their total irrigated area under micro-systems. South Africa is the only other nation reporting more than 10% of its irrigated area under micro-irrigation.

Table A 1.1   Area (ha) Under Micro-Irrigation by Country, Ranked by Rate of Increase, 1986 – 1991

Country	1981	1986	1991
Japan		1,400	57,098
Thailand		3,660	41,150
Austria		220	2,000
Mexico	2,000	12,684	60,600
Italy	10,300	21,700	78,600
Poland		1,522	4,000
Australia	20,050	58,758	147,011
Cyprus	6,600	10,000	25,000
China	8,040	10,000	19,000
Morocco	3,600	5,825	9,766
USA	185,300	392,000	606,000
Hungary	2,500	2,450	3,709
Spain		112,500	160,000
South Africa	44,000	102,250	144,000
UK	3,150	4,690	5,510
Yugoslavia		3,820	3,820
Germany	845	1,850	1,850
Netherlands		3,000	3,000
Chile		8,830	8,830
Portugal		23,565	23,565
Taiwan		10,005	10,005
Jordan	1,020	12,000	12,000
Brazil	2,000	20,150	20,150
Malawi		389	389
France	22,000	50,953	50,953
Ecuador		20	20
Egypt		68,450	68,450
Czechoslovakia	830	2,310	2,310
Israel	81,700	126,810	104,302
Canada	4,935	9,190	6,149
Malaysia		630	177
USSR	11,200		
New Zealand	1,000		
Iran	800		
Senegal	400		
Argentina	300		
Puerto Rico	70		
Tunisia	25		
India	20		55,000
Colombia			29,500
Turkey			32
Philippines			5,041
World	412,710	1,081,631	1,768,987

Source: Bucks (1993).

69

Table A 1.2   Area Under Micro-Irrigation as a Percentage of Total Irrigated Area

Country	Area Under Micro-Irrigation (ha)	As a % of Total Irrigated Area
Cyprus	25,000	71.4
Israel	104,302	48.7
Jordan	12,000	21.2
South Africa	144,000	12.7
Australia	147,011	7.8
Colombia	29,500	5.7
Spain	160,000	4.8
France	50,953	4.8
Italy	78,600	4.7
Portugal	23,563	3.7
USA	606,000	3
Egypt	68,450	2.6
Taiwan	10,005	2.4
Japan	57,098	1.8
Mexico	60,600	1.2
Thailand	41,150	1
Morocco	9,766	0.8
Brazil	20,150	0.7
China	19,000	0.1
India	55,000	0.1
Other	46,837	
TOTAL	1,768,985	

Source: Bucks (1993).

Appendix 2

The Use of Modern Irrigation Technologies by Small Farmers: A Review of Experience

Appendix 2   The Use of Modern Irrigation Technologies by Smallholders: A Review of Experience

Israel

Melamed (1989) and Van Tùijl (1993) describe the development and current status of modern irrigation technology, at a national level in Israel.

In 1948 Israel had 30,000 ha of gravity irrigation systems. By 1990 the area under irrigation was 231,000 ha, using sprinkler and micro-irrigation technologies. The reasons for this rapid transition from surface systems to modern technologies are given as:

- Scarcity and cost of water
- Effective national controls over water allocation and pricing policy
- Well-educated farmers open to innovation. Many were settlers with no tradition of surface irrigated agriculture
- Government support offered to national manufacturers of irrigation hardware
- Well-trained irrigation extension and advisory service supported by irrigation agronomy research.

The water scarcity[3] faced by Israel was apparent to planners from the formation of the modern state of Israel. As a consequence, effective and well-resourced national agencies, capable of overseeing the development of a national water grid and promoting efficient water use in agriculture and other sectors, were established.

The chronology and scope of the measures implemented with central government support are illustrated in Figure A2.1. A Master Plan for the water sector, establishing a Water Commission with legal powers to control water use, grant and revoke water user licences and set water rates, was drawn up in 1951. The Irrigation and Soil Field Service (ISFS) was established in the mid-1950s, under the authority of the Water Commission, to deal exclusively with irrigation extension. In 1965 the Israel Center of Water Works Equipment (ICEW) was set up to promote development of higher efficiency water use devices and to set equipment standards.

Government policy, implemented through the Water Commission, has continually promoted improved water use efficiency in the agricultural sector. Research, development and early manufacture of localised systems by Israeli manufacturers were supported by financial guarantees, (early to mid-1960s). Subsequently, (early 1970s) a national programme of field trails and demonstrations was funded by the Water Commission to promote the use of newly developed micro-irrigation systems and improve the management of sprinkler irrigation. During this time, soft loans and grants were available to farmers and settlements wishing to invest in new technologies. Van Tuijl (1993), reports that it was easier to teach unskilled farmers (i.e. with no 'irrigation culture' or background) how to manage sprinkler or drip systems than to train them in efficient surface irrigation practices.

---

[3] Israel's water availability was estimated to be 473 m^3/head/year, in 1990, decreasing to 307 m^3/head/year by 2025, (Van Tuijl, 1993).

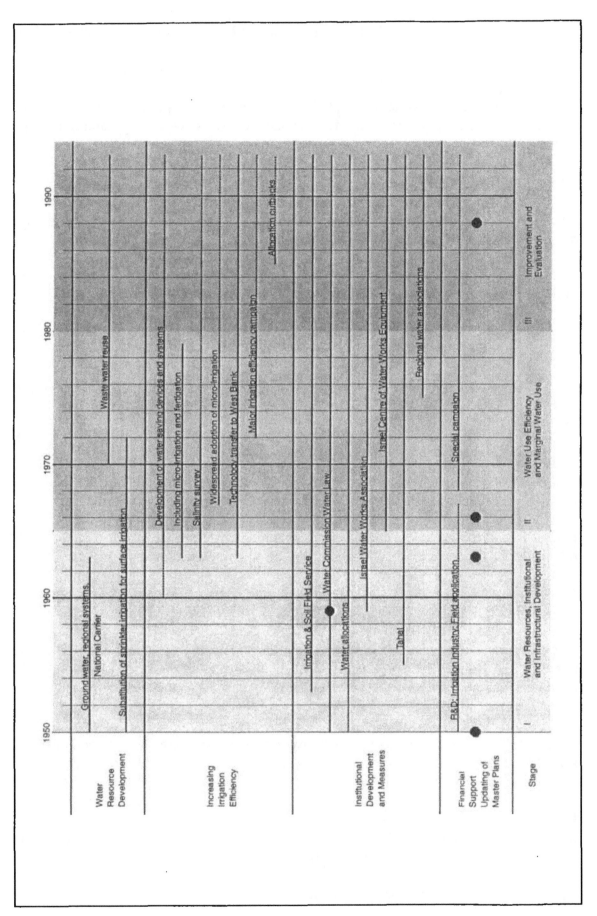

Figure A2.1 Development Stages in Increasing Water Use Efficiency in Israel. Source: Van Tuijl (1993)

The evolution and adoption of modern irrigation technologies in Israel by crop type, is summarised below:

Truck Crops (vegetables)
1950 - 1960: Sprinkler systems adopted. Solid set replacing periodic move to reduce labour costs. Some growers reverted to surface methods to avoid crop damage through wetting of foliage.

Early 1960s: Experimentation with drip laterals to reduce evaporation losses and assess use of brackish water on marginal (desert) soils. Results showed some water saving when compared with sprinkler, but improvements in crop yields and the ability to raise crops on desert soils were more important.

Mid-1960s: Rapid uptake of drip irrigation for vegetable production. Attempts to use portable laterals to reduce costs/ha were abandoned due to high labour costs.

Late 1960s and 1970s: Solid set mini-sprinkler systems introduced to replace earlier solid set sprinklers for close-spaced vegetable crops.

1990s: Almost all vegetable crops with row spacings of 0.8m and greater are irrigated with drip systems. Vegetables with narrower row spacing are irrigated with mini-sprinklers.

Field Crops
Early 1950s: Imported, hand-move sprinkler systems used in place of surface irrigation methods.

Early 1960s: Tractor-drawn, end-tow laterals introduced to reduce labour costs of hand-move systems.

Mid-1960s: Trials of drip systems, originally developed for orchard and vegetable crops.

1970s: Pressure-compensating emitters allow longer lateral lengths and cost savings by reducing the number of secondaries. Mechanised reel systems for laying out and retrieving laterals developed, together with thinner walled pipe, reducing labour and material costs.

1990s: 40% of field crop area under drip irrigation, the remainder under different sprinkler systems. Centre pivots and irrigation booms are reported to be 'gaining in popularity.'

Orchards
Solid set drip and mini-sprinkler systems now irrigate 90% of orchard crops. The remaining 10% are irrigated with under-tree sprinklers on hand drawn, plastic laterals.

Approximate equipment costs are given in Table A 2.1

Or (1985), Regev *et al* (1990) and Keen (1991) describe the introduction of small-scale, drip irrigation systems amongst Bedouin farmers on the west bank of the Jordan Valley in the Jiftlik region. Prior to system modernisation and the introduction of drip technology, water was conveyed in a lined main canal, average flow 0.5 -0.7 m^3/s, and distributed to four villages in unlined distribution channels. Only one-third of the cropped area was given over to vegetables, irrigated in zig-zag furrows, the remaining two-thirds growing traditional varieties of sorghum and barley. Field application efficiencies were about 60% (Regev *et al*, 1990).

Modernisation of the irrigation system included:

- Lining of the four distribution canals
- Construction of 60 farm reservoirs
- Installation of a standard set of irrigation equipment - diesel pump, filtration and fertilisation systems, aluminium distribution mains and drip laterals

Or and Keen stress that introduction of pressurised drip irrigation was only one component of a larger package which included:

- Improved seed varieties
- Use of plastic mulches and low tunnels and solar soil sterilisation
- Increased use of fertiliser, herbicides and pesticides

The first trials and demonstrations of the new 'package' were carried out by the extension service in the early 1970s. Equipment suppliers and foreign NGO funding for loans to farmers then continued the programme. Over approximately 10 years, the package of measures was introduced to 1,600 ha of the original command area. By 1982 95% of the irrigated land was under drip rather than surface systems.

Regev *et al* (1990) calculate the capital cost of equipment, including the cost of canal lining and farm storage ponds to be $US 4700/ha at 1984 prices. They stress that the shift to intensive winter vegetable production and the marketing of produce to well-developed local markets was an essential component of the modernisation package to achieve the needed returns on investment.

Table A 2.1   Average On-Farm Irrigation System Costs in Israel

Crop and Irrigation method	Cost $ US / ha*
<u>Vegetable crops</u>	
Hand-move sprinkler laterals	1,400
Solid set sprinkler laterals	5,700*
Drip – solid set	3,000
Mini-sprinklers - solid set	3,100
<u>Field Crops</u>	
Hand-move sprinkler laterals	1,000
End-tow laterals	1,600
Mechanical move laterals	1,600
Drip – seasonally solid	2,500
Drip – seasonally solid, thin walled	1,300
<u>Orchards</u>	
Hand-move sprinkler laterals	1,600
Sprinklers on plastic drag lines	2,000
Overtree sprinklers – solid set	3,200
Drip – solid set	1,500
Mini-sprinklers - solid set	2,200
<u>Grapevines</u>	
Drip	2,200

* The costs, at mid-1988 prices, exclude mains and pumping plant except in the case of solid set sprinklers.
Source: Melamed (1989).

Cyprus

Water availability per capita in Cyprus is somewhat greater than in Israel and Jordan, the other two Mediterranean countries irrigating a very high percentage of their total irrigated crop area with modern technologies. Van Tuijl (1993) estimates 1189 m^3/head/yr available in the year 2000, compared with 404 m^3/head/yr (Israel) and 200 m^3/head/yr (Jordan).

Cyprus has a total irrigated area of 55,000 ha, of which 27,000 ha (49%) are irrigated using sprinkler and micro-irrigation technologies. Approximately half the total irrigated area (28,200 ha) has been developed under major public schemes, the largest being the Southern Conveyor Project (SCP) which provides a pressurised pipe distribution network serving 13,450 ha (Van Tuijl, 1989 and 1993). The SCP is a multi-purpose project providing domestic water supply to the major population centres of Cyprus as well as developing new irrigation areas. Allocating half of the total project cost to irrigation, Van Tuijl (1989) reports the cost per irrigated hectare as $US 12,300.

Farms benefiting from these public schemes are small, with fragmented holdings totalling less than 1 ha. Land consolidation has been a key element in the design of the pressurised distribution networks down to the farm level. Water from the main pipeline is diverted into storage reservoirs at night when municipal demand is minimal. Each reservoir serves an area of about 400 ha. From the reservoir, water is piped to hydrants serving up to 30 ha. Pumps are used where the gravity head is not sufficient to give adequate operating pressure at the hydrant outlet. A single hydrant may have up to four outlets, each serving three sub-units of 2.5 ha. Each sub-unit is made up of a maximum of three farm plots.

Water meters, pressure and flow regulators and filters were installed at each outlet, potentially serving up to 9 farmers. Water is now metered at the turnout to farmer's plots to facilitate individual farmer billing.

The first hand-moved sprinkler systems were imported into Cyprus in 1965 and drip systems were first imported from Israel in 1970. To encourage the adoption of these new 'water-saving technologies' the Ministry of Agriculture and Natural Resources (MANR), under the Water Use Improvement Project of 1965 provided a subsidy of 15% of purchase and installation cost, with the balance provided as a loan at a low rate of interest.

Local manufacture of equipment is now well established in Cyprus.

Jordan

Battikha and Abu-Mohammad (1994) report the methods of irrigation and areas under command in three principal zones in Jordan.

Table A 2.2   Areas Under Different Irrigation Types in Jordan 1990/91 (ha)

Location	Surface	Sprinkler	Drip	Total
Jordan Valley	13,400 (47%)	200 (0.7%)	15,000 (52%)	28,600
Southern Ghors	1,600 (42%)	--	2,200 (58%)	3,800
Highlands	5,700 (18%)	5,100 (16%)	21,100 (66%)	31,900
Total	20,700 (32%)	5,300 (8%)	38,300 (60%)	64,300

Irrigation in the Jordan Valley and Southern Ghors has been promoted by the construction of major canal infrastructure to capture and convey surface water from wadi systems to agricultural areas. In contrast, the development of irrigation in the Highlands, which represents almost half of Jordan's irrigated agricultural area, has proceeded without the creation of any central public authority, infrastructural project or wide-scale extension effort. Farmers in the highlands pump groundwater from aquifers as deep as 400 m, adopting pressurised irrigation systems to maximise water use efficiency in the face of high pumping costs.

Construction of the East Ghor Canal Project, in the Jordan Valley, began in 1960. Stage I of the project, completed in 1969, provided surface irrigation to about 13,000 ha (Van Tuijl, 1993). A major land consolidation programme with legislation to prevent future fragmentation of holdings was an important component of this first stage. The irrigated farm holding is set at between 3 and 4 ha, depending on soil type.

Stage II, beginning in 1973, focused on broader regional development objectives to raise standards of living in the rural areas and reduce rural migration. Investment was made in roads, housing, schools, and health centres together with agricultural processing and marketing centres. During this time, pressurised pipe networks to convey water from the main canal to farms, exploiting natural land slope to provide pressure at farm turnouts, were developed. They were installed, by the state agency, as part of a programme to introduce sprinkler irrigation. However, farmers became interested in drip irrigation systems, encouraged by the private sector, and the agency's plans for modernisation were overtaken by events.

Or (1993) and Van Tuijl (1993) report that sprinkler systems, purchased by the Jordan Valley Commission and delivered in 1978, were almost obsolete on arrival. Private sector companies began introducing drip systems in 1975, establishing demonstration plots and providing advisory services to farmers. Dealerships and commercial banks provided credit for equipment purchase when parastatal credit agencies refused to provide credit for 'unproven technologies' (Van Tuijl, 1993). Installation costs were about $US 3,600 / ha. Adoption was rapid and farmers paid back credit in 2 to 3 years. More recently drip equipment has been manufactured in Jordan, and costs have fallen to about $US 1,000 / ha.

An important element in the rapid adoption of drip irrigation systems was a strong and profitable export market for winter vegetables to surrounding Gulf States. Drip systems are used mainly for vegetable production - tomatoes, aubergines, cucumbers, onions, chillies. Micro-sprinklers and bubbler systems, introduced more recently, have been used in perennial crops - citrus, banana, peach, apples. Van Tuijl (1993) reports that average vegetable yields increased from 8.3 t/ha in 1973 to 18.2 t/ha in 1986. In the same period fruit yields increased from 7.1 to 16.0 t/ha. The yield increases occurred in response to a package of improved agricultural practices and are not solely the result of changes in irrigation method.

The Southern Ghors project, implemented by the Jordan Valley Authority (JVA), developed irrigation in the low-lying land to the south of the Dead Sea. Land consolidation was an important component of the project and Hanbali et al, (1987) emphasise the role of the JVA as a central government agency, empowered to redistribute lands.

Construction of intakes on six wadis, with settling basins and storage ponds supplying distribution pipelines, was completed in 1985. High capacity, sand media filters are operated and maintained by the JVA to provide primary filtration of water. Each farm turnout is equipped with a gate valve, flow limiting valve and water meter, controlled and operated by the JVA. Below this, a gate valve and secondary, mesh filter, are operated and maintained by the farmer.

India

According to Keller (1990), there is a 'ferment of interest' in modern irrigation technologies in India but, 'out of several thousand installations, few are being well maintained and operating satisfactorily and many have even been vandalised'. Information on the extent of pressurised irrigation systems in India is often conflicting, but estimates from the literature are presented in Tables A 2.3 and A 2.4.

Current irrigated area is given as 62 mha by Saksena (1995). It is estimated that available surface and groundwater resources could irrigate a maximum of 113 mha under surface irrigation (Singh et al, 1993). Annual food production is currently 160 million tonnes of grain equivalent, for a population of 900 million, (Suryawanshi, 1995; Saksena, 1995). It is estimated that the grain equivalent required by 2000 will be 240 million tonnes. To meet the shortfall in production, the productivity of existing irrigated lands must be raised. Several authors suggest that the widespread adoption of drip and other micro-irrigation technologies can contribute significantly to this need.

There are important weaknesses to this argument, the most fundamental being that drip systems are not appropriate for the irrigation of staple grain crops but are used universally for the irrigation of higher value cash crops. Sharma & Abrol (1993) acknowledge that drip must be targeted at selected environments where water costs are high, soils, topography and/or water quality make surface irrigation impractical, and high value cash crops can be grown and marketed.

Water availability in India varies widely between states as shown in Table A 2.5. Sharma and Abrol (1993), state that groundwater supplies approximately 50% of the net irrigated area in India, and provide information on the increasing number of private and public pump sets in use, (Table A 2.6). Over-exploitation of groundwater has led to falling aquifer levels in Punjab, Haryana, Western Uttah Pradesh and Rajasthan.

States where the greatest development of modern irrigation systems has taken place are those where water is most scarce: Maharashtra, Karnataka and Tamil Nadu (Chauhan, 1995).

Imported drip systems were first evaluated in the early 1970s. Tamil Nadu Agricultural University, Coimbatore, conducted in-field demonstrations but the equipment failed frequently and farmers consequently showed little interest. The first National Seminar on Drip Irrigation took place at Coimbatore in 1981. In the same year the National Committee on Use of Plastics in Agriculture (NCPA) was established under the Ministry of Petroleum. The NCPA aims to promote the use of plastics in agriculture and is responsible for 23 Plasticulture Development Centres, (Chauhan, 1995).

Central and state governments have offered subsidies to small and marginal farmers since 1983, to encourage uptake of drip systems (Singh et al, 1993). The value of the subsidies varies from state to state and depends on the farmer's land area and form of tenure. Saksena (1993b) and others, criticise the targeting of subsidies exclusively to small farmers, arguing that the technology would be promoted more effectively by encouraging larger, more progressive farmers, to invest in the systems and smaller farmers would then see the benefits.

There are approximately 50 companies manufacturing and promoting drip technologies in India (Chauhan, 1995), of which Jain Irrigation Systems Ltd (JIS) is one of the largest, having a licence agreement with a major US manufacturer to manufacture and market their products in India. Suryawanshi (1995), of JIS, describes the role of JIS in promoting drip irrigation equipment in Maharashtra state.

- The company imported equipment and evaluated it on its own farm, demonstrating the trials to local 'progressive farmers'. Drip systems were then installed on the farms of interested farmers with the company paying all costs. It was agreed that farmers would pay for the equipment only if higher incomes were obtained from improved yields.
- JIS carried out village demonstrations and field visits for farmers and at the same time encouraged government to provide funds for farmer subsidies and credit.
- Working with a US manufacturer, JIS have developed low-cost, 'simple' components and field designs, recognising that many farm plots are less than 1 ha.
- The company negotiated with national government to secure funding for subsidies and soft loans for smaller farmers.

The following support is offered to farmers wanting to install systems:

- Field survey and collection of data on climate, soil, water
- Soil and water analysis
- System design
- Assistance in securing subsidies and loans
- Delivery of equipment to the farm and field installation
- System commissioning and training of farmers

JIS also offers two free after-sales services of the equipment in the first 6 months and continued monitoring of system performance and advice, through periodic visits.

Singh *et al* (1993) report that Maharashtra State, where JIS are based, include 66% of the total area under drip irrigation in India in 1993[4]. Several factors, including severe water scarcity, crop types, state subsidies and marketing opportunities, have led to the relatively rapid expansion of drip irrigation in this state. However, the activities of JIS as a private sector company, are a significant factor in the expansion of drip systems.

---

[4] Note that the 31,300 ha under drip Maharashtra State are equivalent to 82% of the entire area of drip irrigation in Jordan. The relative 'success' of modern irrigation technologies within countries is often gauged by the fraction of total irrigation under modern systems. However, it should be noted that neither Jordan nor Israel irrigate significant areas of basic food grains. In making comparisons between countries and regions differences in production must be recognised.

Table A2.3   Estimated Areas under Drip Irrigation in India (ha)

Source	1981	1987	1988	1989	1990	1991	1992	1993	1994	1995
Bucks (1993)	20					55,000				
Singh et al[1] (1993)		250	1,680	4,100	8,670	14,420	29,000	43,680 (47,300)[2]		
Saksena			23,500[3]			35,000[4]			70,000[5]	
Suryawanshi (1995)										50,000

Notes:

1. Data show areas receiving state subsidies and therefore exclude larger farmers and commercial plantations
2. Includes an estimate of the area operated by larger farmers and commercial estates not receiving state subsides
3. Saksena (1992)
4. Saksena (1993a) Based on data from National Bank for Agriculture and Rural Development and figures from national manufacturers.
5. Saksena (1995) cites data from Indian National Committee on Irrigation and Drainage, published in July 1994.

Table A 2.4   Estimated Area Under Sprinkler Irrigation in India (ha)

Year	No of Sprinkler Sets	Irrigated Area (ha)
1989	11,400	58,000
1991*	17,200	76,800
1995*	28,000	116,800
2000*	44,800	200,000

Source: Sharma and Abrol (1993)
* Data for these years are projections based on past growth

Table A 2.5   Water Availability by State (m³/head/year)

State	1981	2001
Maharashtra	788	532
Tamil Nadu	820	554
Kerala	860	561
West Bengal	1,068	722
Karnataka	1,099	743
Gujarat	1,206	815
Uttar Pradesh	1,246	842
Rajasthan	1,261	851
All India	1,353	910
Bihar	1,359	918
Jammu & Kashmir	1,739	1,175
Madhya Pradesh	1,976	1,335
Assam	2,136	1,442
Andra Pradesh	2,230	1,506
Orrisa	2,244	1,517
Punjab	3,450	2,333

Source: Singh *et al* (1993)

Table A 2.6   Number of Electric and Diesel Pumpsets for Tube-wells in India

Plan period	No. of pump sets ('000)
1951	21.0
1961	198.9
1969	1,088.8
1974	2,428.2
1980	3,965.8
1990	8,226.2
Diesel pump sets	4,550.0
Total	12,776.2

Source: Shrama and Abrol (1993)

Data on yield and water saving benefits and installation costs for different crops are presented in Tables A 2.7 and A 2.8. The data were obtained from trial plots operated by universities or progressive farmers with supervision from equipment suppliers, and indicate the potential water savings and financial benefits obtained in converting from surface irrigation methods to drip. Saksena concludes that:

"In inspite of so many advantages [that] the micro-irrigation system possesses and the subsidy given by government and the loan facility available from banks, the system has not made much progress and headway. Farmers are very slow adopting this and only in the case of horticulture and cash crops. The progress is rather uneven and slow." (Saksena, 1993b).

A reconnaissance survey carried out by the Water and Land Management Institute (WALMI) of Maharashtra (Holsambre, 1995) shows that the design, performance and maintenance of many drip systems is well below potential. The survey examined systems on 12 farms selected at random across the state. Emission uniformity was 85% or better on four of the farms but values of 50% or less were recorded in half the sample.

The main cause of poor performance was the mismatch of pumpsets to the head/discharge requirements of the drip systems. High head, low discharge pumps are needed. In the majority of cases farmers were using low head, high discharge pumps normally used for surface irrigation. Only one of the systems had a sand filter, although the water source on six of the farms contained algae and mud where sand filtration is recommended. Three of the farms had neither sand nor screen filters and emission uniformity was reduced to between 35 and 50%. Eight of the farms studied were on steep or rolling terrain. A major weakness in system design was the failure to take account of variation in head due to the slope of the land.

Table A 2.7   Indicative Installation Costs of Drip Irrigation Systems, Maharashtra State

Crop	Cost /ha (US $)	Expected System Duration (years)
Sugarcane	1,300	7
Banana	1,400	10
Tomato	1,300	7
Sweet lime	800	10
Betelvine	1,300	10

Source: Suryawanshi (1993)
Includes cost of drip laterals, secondaries, main and filter. Excludes cost of pump and borehole.

Dalvi *et al* (1995) report on a survey of 42 drip systems in Maharashtra State carried out in 1990. Only 17% of the systems had distribution uniformities of 90% or greater. Inadequate filtration - leading to blocked emitters - and leakage of pipes at joints, were the major causes of low distribution uniformity. System layout and pipe sizing was generally acceptable, although savings could have been made by using smaller sub-mains on 15 of the systems. The mismatch of pump characteristics to system requirements was again a common problem. All farmers reported some degree of water saving but less than a quarter of the sample reported any improvement in yields when

83

compared with surface methods of irrigation. The survey lists the following constraints identified by farmers regarding the purchase and operation of drip systems.

Table A 2.8   Yield Increases and Water Saving Under Drip Irrigation

Crop	Yield (Tonnes/ha)			Water Use (mm)		
	Conventional	Drip	% yield increase	Conventional	Drip	Water Saving (%)
Banana	57.5	87.5	52	1,760	970	45
Grapes	26.4	32.5	23	532	278	48
Sweet lime	100	150	50	1,660	640	61
Pomegranate	55	109	98	1,440	785	45
Papaya	13.4	23.5	75	228	74	68
Tomato	32.0	48.0	50	300	184	39
Watermelon	24.0	45.0	88	330	210	36
Ocra	15.3	17.7	16	54	32	40
Cabbage	19.6	20.0	2	66	27	60
Chillies	4.2	6.1	44	110	42	62
Sweet Potato	4.2	5.9	39	63	25	60
Beetroot	46	49	7	86	18	79
Raddish	70.0	72	2	46	11	77
Sugar Cane	128	170	33	2,150	940	56
Cotton	2.3	2.9	26	90	42	53

Source: Singh *et al* (1993)

Table A 2.9   Problems Reported by Farmers Operating Drip Systems

Problem	Constraint	% of farmers responding
High cost of spare parts	Financial	88
High initial cost	Financial	66
Difficult to operate	Technical	62
Scheduling of irrigation unknown	Knowledge	52
No knowledge of chemicals to remove clogging	Knowledge	48
Analysis of soil & water not carried out	Knowledge	38
Emitter clogging	Technical	35
Leakage at dripper/lateral joint	Technical	33
Pipes damaged by rodents	Technical	28
Broken emitters	Technical	26
Leakage at filter/main joint	Technical	19
Method of measuring pressure/discharge not known	Knowledge	16
Incorrect filter unit	Technical	11
Leakage at lateral/sub-main Joint	Technical	11
Pipes damaged by birds	Technical	11
Theft of pipes	Technical	11
Pipe damage by Implements	Technical	9

Shelke *et al* (1993) surveyed farmers using sprinkler irrigation in the Sikar district of Rajasthan, where it is estimated that approximately 4000 sprinkler sets have been sold. Of the 40 farmers in the survey, 52% were classified as illiterate, suggesting that the technology was not only being adopted by well-educated farmers. 70% of the farmers were irrigating holdings of 3 ha or less, 32% owning less than 2 ha. The recommended operating pressure was in the range 200 - 400 kPa but 60% of farmers were operating their systems at pressures below 100 kPa due to inappropriate pump sets. Incorrect operating pressure, and highly variable lateral spacing, led to low application uniformity on most of the farms. Despite these shortcomings, the study showed that farmers with the smallest landholdings of less than 2 ha, achieved a benefit: cost ratio of two. Farmers liked the sprinkler systems as they were considered to save labour and allow unlevelled land to be irrigated.

Rao (1992) describes the use of sprinkler irrigation by a co-operative of 16 'marginal' farmers in the Eastern Dry Zone of Karnataka state where irrigation depends on groundwater sources. Driven by the failure of traditional shallow wells and heavy state subsidies on electricity for pumping, the number of tube-wells in the state increased five-fold to 41,000 in the five years up to 1987. The wells, equipped with electric submersible pumps, yield between 2 to 5 l/s and cost about US$2,000 (1986 prices).

Farmers in the co-operative each had holdings of less than 1 ha, their combined land area being 13.2 ha. The co-operative received a state grant for the full cost of three tube-wells, pump sets and sprinkler systems. Each farmer retained ownership of his own land and was free to choose his own cropping pattern but the quantity of water allocated to each was independent of crop type or size of holding. 75% of the area is planted to perennial crops - mulberry, coconut, mango and banana. Ragi is grown on 3.2 ha in the summer monsoon, and vegetables in the winter.

Rao makes little comment on the appropriateness of the sprinkler technology for these farmers but he refers to pump sets frequently burning out due to voltage fluctuations. He suggests that the co-operative is weak and concludes that further support from extension services is essential to secure the widespread adoption of sprinkler systems by marginal farmers.

China

Estimates of the area irrigated by modern methods vary widely. Kezong (1993) gives an overview of 'water saving irrigation techniques' promoted by means of low interest government loans repayable over 2 to 5 years. He defines the following irrigated areas and system types:

Sprinkler:
- 53,000 ha around Beijing. Buried PVC mains carry water from tube wells to field hydrants. Aluminium portable laterals are used in the field.
- 11,500 ha, predominantly solid set systems, near Shanghai, irrigating vegetables.
- Hunan Province - 5,000 ha citrus and tea orchards on sloping land. Permanent, solid set systems.
- Xinjiang Autonomous Region - 14,000 ha gravity-head sprinkler systems

Low Pressure Pipelines:
- Promoted in northern China to improve conveyance efficiencies from tube-wells. Buried pipes supply hydrants at approximately 50 m spacing. Pipes may be of thin walled PVC or spun concrete. Lay-flat hose may be used to convey water from hydrants to individual basins. It is estimated that approximately 2.5 million ha are irrigated by pipeline systems.

Micro-irrigation:

- Approximately 20,000 ha of vegetables and orchard crops.

Chen Dadiao (1988) reports that 650,000 ha were under sprinkler irrigation in 1984. He describes small, portable sprinkler machines, made locally, comprising a small petrol engine and pump delivering water to a single rain-gun mounted on the pump frame. This is the most widespread type of sprinkler system.

Backhurst (1995) describes a pilot project, with funding from the European Commission, to grow horticultural crops under sprinkler irrigation on sandy wasteland soils of Jiangxi province in southern sub-tropical China. The pilot area of 210 ha is operated in 2 ha blocks managed by individual farmers. Portable sprinkler systems are being used.

Yin (1991) gives more information on the use of layflat hose in northern China. Fully movable systems have all pipe-work above ground and semi-fixed systems consist of a buried pipe system supplying hydrants to which layflat hose is attached. The buried pipe network consists of layflat hose protected by an outer shell of cement mortar. Layflat hose was first introduced in 1979 and is reported to be used on at least 2 million ha.

Qiu (1992) describes the development of drip irrigation equipment by a subsidiary of the Institute of Water Conservancy and Hydroelectric Power Research, Beijing. Imported systems are estimated to cost up to $US 4,000 / ha. Local systems use micro-tubing emitters to reduce cost and problems of clogging. Qiu states that the equipment has been used successfully to irrigate small areas of field grain crops by moving laterals between sets.

Pakistan

There has been very limited application of modern irrigation technologies to date in Pakistan. Keller and Burt (1975) made an early study of the potential for introducing trickle and sprinkler irrigation in Pakistan in 1975. It was concluded that conventional drip systems would be inappropriate due to cost and the need for skilled system management but evaluation of hose-basin systems and hose pull under-tree sprinklers systems was recommended. National manufacturers, capable of producing the required hoses and sprinkler fittings, were identified. The systems were for application in:

- The peri-urban area around Karachi
- Northern and western parts of the Thal desert where groundwater is available
- Fringe areas around existing surface-irrigated areas

The systems would be used primarily to irrigate tree crops - orange, apple, mango and nuts - and some vegetables.

Methods of improving on-farm water use were evaluated, fifteen years later, under the USAID funded Irrigation System Management - Research project (ISM/R, 1993).

86

Improved surface irrigation methods using level borders and furrow methods were evaluated and the status of sprinkler and drip irrigation technologies reported.

Portable rain-gun sprinkler systems, similar to those marketed in China, are manufactured in Lahore. Equipment is estimated to cost $US 500 / ha for systems using diesel pumps and $US 300 / ha where an electric pump is used. Eight standard sets are marketed for farm areas from 0.8 to 20 ha.

Moshabbir *et al* (1993) describe the work of the Pakistan agricultural Research Council (PARC) to promote national production of drip irrigation components. Low-density polyethylene pipe for laterals, spiral flow emitters and micro-tubing, are manufactured in Lahore. Equipment is estimated to cost approximately $US 800 / ha, excluding the cost of the pump and main line.

Guatemala

Irrigation by gravity-driven sprinklers has been successfully introduced in hillside farming systems. The programme began in 1978 with technical assistance from USAID, working with the agricultural extension service. Schemes range from 5 to 60 ha. Individual farmers irrigate plots from 0.2 to 1.4 ha (Lebaron, 1993). Two hundred and fifty village systems had been established by 1989.

Villagers wanting to establish a small-scale irrigation project approach an extension service team comprising an engineer, agronomist, surveyor and draughtsman. In order to secure a loan for equipment, villagers must form a water user association. Loans are repaid over 20 years at interest rates as low as 2% (Lebaron *et al*, 1987). Technical assistance for system design and installation is provided free.

PVC pipes convey water from hill springs to standpipes provided for individual farm plots. Sprinklers are attached to the standpipes by lengths of garden hose (15 m length, 16 or 20 mm dia.) A single sprinkler is assumed to irrigate an area of approximately 600 m^2 in nine moves. Earlier systems were designed to operate 'on-demand', but to reduce the cost of supply pipes, the design of systems was subsequently based on rotational supply, managed by the user group.

Equipment costs vary from $US 150 to $US 2400 / ha (1986 prices). When the costs of labour and technical assistance are included these figures are $US 320 to $US 6800 /ha. Even the most expensive of these schemes was estimated to yield an internal rate of return of 9% (Lebaron ,1993).

Farmers are encouraged by the extension service to grow high value cash crops such as strawberries, flowers, potatoes and other vegetables, in the dry season, replacing traditional crops of maize and beans. Where farmers have changed to these higher value crops the returns are very high. Lebaron (1993) reports that even where farmers choose to irrigate traditional crops positive benefits are still achieved because of the low cost of the loans and low operating costs.

Support and technical assistance regarding system design and installation was good, but Lebarron *et al* (1987) report that there was little formal advice on crop agronomy and water management.   Despite this absence of formal extension information, diversification into higher value, dry season vegetables, has been widespread.

Sri Lanka

Small farmers in Sri Lanka have not, to date, adopted modern irrigation systems though some equipment trials have been carried out.

A low head drip (LHD) system, constructed from locally available materials, was evaluated in 1988 (Miller and Tillson, 1989). The system operated at a variable gravity head of between 2.0 and 0.5 m and irrigated a plot of 1 ha. Mains and manifold fittings were of lightweight, 110mm PVC to which layflat pipe of different types, was connected.   Commercial drip laterals were connected to the manifolds using two different connectors. The systems were operated for one growing season - April to July 1988 - by labourers with no previous experience of drip irrigation. Despite the apparent technical success of the trial there is no evidence of the system being promoted by agencies in Sri Lanka. Batchelor *et al* (1993) state that cost and the need for relatively skilled management and better crop husbandry were factors working against the adoption of this LHD system. Furthermore, they conclude that the 1 ha system irrigated too large an area. Many farmers farm plots of less than 1 ha and the absence of control valves made it difficult to vary the irrigation frequencies and depths applied to different sections.

Smaller and cheaper LHD systems have been evaluated in Zimbabwe.

De Silva (1995) used locally made drip irrigation equipment to irrigate a 2 ha plot of aubergines, chillies and onions in a study of shallow-dug wells in North West Province, but gives no data on the cost or technical performance of the system.

Zimbabwe

Watermeyer (1986) states that, up to 1986, Communal Land Sprinkler systems had been "complete failures" due to a failure to implement schemes in consultation with small farmers and inadequate user training. He concluded that effective operation required an agreed crop rotation (crop mix), rigid adherence to an agreed cropping calendar (timing of operations) including the timing of lateral moves, and establishment of a maintenance fund.

Despite this early criticism, The Department of Agriculture, Technical and Extension Services (AGRITEX) of the Ministry of Lands and Rural Development, has gone on to design and install several draghose sprinkler systems on Communal Lands with apparent success.

Solomon and Zoldoske (1994) report that the quality of locally manufactured sprinkler heads is particularly poor, with nozzle variation resulting in highly variable discharge between sprinklers of the same type. They also observed that media filters used in drip installations by commercial farmers were often substantially undersized and poorly maintained.

Three methods of sub-surface irrigation, appropriate for use in small community vegetable gardens, have been evaluated in the semi-arid area of southeast Zimbabwe (Batchelor *et al*, 1996). The vegetable gardens are small, with a total cropped area of only 0.5 ha, and individual holdings vary from 50 to 100 m^2. Water is hand pumped from shallow, hand-dug, large-diameter wells. The yield of these wells is improved by drilling horizontal laterals out from the base of the well to a distance of up to 30 m to create a collector well (Lovell *et al*, 1996). The area of these gardens is small, which limits the contribution they can make to the national food production. However, Table A2.10 shows that the annual gross margins generated by the gardens, on a per hectare basis, compare very favourably with other irrigation schemes in Zimbabwe, establishment costs are similar, or slightly higher, and the calculated IRR is higher than on several other types of scheme.

The study concluded that the system, despite its simple character, was still seen as expensive. The omission of filtration - to reduce costs - was reported as a major problem, as emitters had to be cleaned regularly. The sub-surface clay pipe system was the "best practical alternative to flood irrigation for irrigating vegetable plots that are, say, 0.01 to 1 ha in area" (Batchelor *et al*, 1996). Lovell *et al* (1996) report that the clay pipe technique has been adopted by "some gardens in the region", and where they are used the time spent in watering has fallen from 20 to 5 hours per week.

Table A 2.10   Indicative Values of Agro-economic Performance of Various Scales of Irrigation System Operating in Southern Zimbabwe. After Lovell et al, 1996

Name	Size (ha)	Type of scheme	Number of members	Average area per family (ha)	Annual gross margin (US$/ha)	Gross margin per unit of water (US$/ha)	Typical cost per hectare (US$/ha)	IRR (%)
ADA Chisumbanje[1]	2,400	River water pumped to canals & syphons	118	3.6	191	N/a	322[2]	5
AGRITEX Towona[1]	151		245	1.2	264	201	3,220[2]	8
AGRITEX Mabodza[1]	12	Gravity fed from dam to canals & syphons	92	0.13	399	217	9,200[2]	3
AGRITEX Chirogwe[3]	5		105	0.05	725	N/a	5,520[3]	13
DAMBO GARDEN Mushimbo[1]	12	Buckets of water from shallow dug wells	14	0.89	237	311	460[4]	52
DAMBO GARDEN Mbiru[5]	4		57	0.07	530	309	460[4]	115
COMMUNITY Romwe[5]		Collector well & two handpumps, water by buckets to community garden	46	0.01	1,832	4,802	8,833[5]	12
Muzondidya[5]			134	0.005	1,675	3,515		11
Gokota[5]	0.5		112	0.005	2,341	3,772		15
Dekeza Sch[5]			49	0.01	*	*	*	*
Mawadze[5]			50	0.01	*	*	*	*

Source:

1) Meinzen-Dick et al (1993) Agro-economic performance of small holder irrigation Zimbabwe, UZ/IFPRI/Agritex Workshop, Zimbabwe, Aug-96.
2) FAO (1994) National Action Programme on Water and Sustainable Agricultural Development, Zimbabwe.
3) Agritex (pers. Comm) Figures based on first two years of operation, 1991-93.
4) Estimate based on cost fencing alone.
5) Lovell et al. (1994) Small scale irrigation using collector wells pilot project Zimbabwe: 4th Progress Report, Institute of Hydrology, UK
6) Financial analysis: IRR calculated for a common project life of 40 years (assuming proper maintenance and sustainable use of natural resources) and a social discount rate of 13 percent.

*) Not yet completed one full year.

www.ingramcontent.com/pod-product-compliance
Lightning Source LLC
Jackson TN
JSHW061959140125
77033JS00050B/609

```
┌─────────────────────────────────────────────────────────────────────────┐
│                         Units of Pressure                                 │
│                                                                           │
│   1 Pascal = 1 Newton/m²                                                  │
│                                                                           │
│   Operating pressures are specified throughout in kiloPascals (kPa)    -  │
│                                                                           │
│   1kPa = 0.145 psi                           1 psi = 6.89 kPa             │
│   1kPa = 0.01 bar                            1 bar = 100 kPa             │
│   1kPa = 0.1 metre water head                                            │
│                                                                           │
└─────────────────────────────────────────────────────────────────────────┘
```

The support of DFID for the two-year study is gratefully acknowledged.

Dr Charles Hutchinson of the Institute of Hydrology is thanked for his advice and assistance.

* 9 7 8 1 8 5 3 3 9 4 5 7 7 *